AF575630

Marvelous Magnetic Machines

Building Model Electric Motors from Scrap

Written and illustrated by

H.P. Friedrichs

Published by Artisan Ideas, 2020.

Written and illustrated by H.P. Friedrichs
Layout and cover design: Andrea James Blaho
Technical Editor: Leonardo Barsantini

ISBN 9781733325042
Library of Congress Control Number: 2020945904
Printed in China.
Artisan Ideas is an imprint of **Artisan North America, Inc.**
Tel: 800-843-9567
Info@ArtisanIdeas.com

Artisan North America
753 Valley Road
Watchung, NJ 07069-6120

Ordering information:
Quantity sales: Special discounts are available on quantity purchases by corporations, associations, and others. For details, contact the publisher at the address above.
For single copies, please try your bookstore first. If unavailable this book can be ordered through: www.ArtisanIdeas.com.

To see our complete selection of books and DVDs on this and other subjects visit our website listed above.

Table of Contents

Introduction

Every year, the city and the surrounding communities where I live host an event known as the "Tucson Gem and Mineral Show". In this event, exhibitors and enthusiasts from around the planet gather to buy, sell, and trade rocks, crystals, gems, jewelry, fossils, and lapidary supplies, among other things. If you're in the market for a huge diamond for an engagement ring, a chunk of genuine meteorite, the tooth of a T. Rex, or a phone-booth-sized amethyst geode, this is the place you want to be.

Readers of my previous books, *The Voice of the Crystal* and *Instruments of Amplification*, can well imagine that such a show is an ideal place to hunt for naturally-occurring semi-conducting materials like galena, pyrites, and other metallic crystals. It was during a hunt for these types of minerals that I came upon a vendor with a large barrel of what he creatively referred to as "rattlesnake eggs". Needless to say, these "eggs" were not eggs at all, but toys. Sold in pairs, each was the size of a horse pill, elliptical in shape, and was fashioned from hematite. The hematite was strongly magnetized, and pairs of such "eggs" would attract each other and stick together aggressively.

To play with the toy, one separates the two eggs, one in each hand, and tosses them gently into the air, such that their flight paths are roughly parallel. The magnetic field in each egg is so strong that if the two of them come within a few inches of each other, they will attract one another and join together in mid-flight. On impact, on account of their shape, the two eggs rock violently back and forth upon each other's surfaces, producing a loud chatter that is reminiscent of a rattlesnake's rattle.

I purchased a set and, while I found them amusing, I was even more delighted by my 2-year-old grandson's enthusiasm for them. In particular, I enjoyed watching his experimentation. Obviously, the two eggs could attract each other, but he soon discovered that if they were oriented a certain way, the eggs would also repel one another. He also noted that whatever the magical force was that existed between these eggs, that force could easily pass through a table top, a magazine, or the palm of his hand.

In watching him, I was reminded of my own childhood and my early interest in magnets. At that time, the best magnets were made from an alloy called Alnico, and a great source of such magnets was the loudspeaker torn out of a discarded transistor radio. The object of my desire was the shiny cylindrical magnet buried inside the speaker's yoke. A vise and a screwdriver, or sometimes just a hammer, was enough to set it free. It was then added to what became a growing collection. As I watched my grandson play, I noted that his discoveries mirrored my own.

Why the fascination with magnets? Mind you, magnetism is not the only force that can influence physical objects over a distance. If you've ever reached into a shipping box and found your sleeve the object of affection of foam packing peanuts, or had your hair stand on end because you pulled a wool sweater over your head, you are familiar with electrostatic forces. And let's not forget gravity, a force that, at least in day-to-day life, is the most evident and dominant force of nature. Yet, large electrostatic fields cannot move the mass that even small magnets can, and the influence of gravity, while potent, can't easily be controlled to suit our purposes. Magnetism has the unique property of being both powerful and manipulable.

Not only can we exploit the attributes of permanent magnets like the "rattlesnake eggs" to both attract and repel, it is possible to produce magnets on demand, by winding coils of copper wire around certain kinds of core materials. Apply electricity, the core becomes magnetic. Remove the electricity, and the magnetism goes away. Apply and remove that current in a cyclical way, and we can build devices that oscillate like buzzers or bell ringers. Arrange the machinery so that the magnetic fields generate torque on a shaft, and you have an electric motor. Motors are a natural next-step in the progression of experimentation and play with magnets. Build a homemade motor that works, and you've demonstrated a certain mastery of a force that, because of its utility, is largely responsible for our present quality of life.

Over the years, I can remember building quite a number of crude motors. One of my favorite authors, Alfred Morgan, in his 1946 classic, *The Boy's Book of Engines, Motors, and Turbines*, describes the construction of a primitive electric motor from bolts, dowel rod, and wire. I've seen at least one 50s era *Boy Scout Handbook* with similar instructions. UNESCO's book, *700 Science Experiments for Everyone*, another classic and a favorite of mine, documents three motor designs featuring raw materials like knitting needles, cork, and test tubes. Countless other books, and now internet web pages as well, offer advice and instructions to help people graduate from simple tricks with "rattlesnake eggs" to building primitive homemade motors. So, it's fair for you to ask what might have possessed me to put together yet another book on this topic.

For my tastes, the motors spawned by most of these plans are very crude. I don't think this an accident, but rather, a matter of purposeful design. An assumption has been made that any higher level of quality would require the use of the sorts of tooling that most experimenters have no access to. I'm speaking, of course, of machine tools like lathes, mills, and such.

I won't argue that nice tools in the hands of a craftsman can produce absolute wonders, but it does not necessarily follow that a lack of tools means we're out of luck. A good dose of creativity and out-of-the box thinking can go a long way toward leveling the playing field. Understand that at its heart, real engineering is not so much about the limitations imposed by the laws of physics and finance; rather, it's the creative manipulation of those limitations to achieve the desired goals, despite them.

To demonstrate this, I set out to build a series of model motors that, at least in my mind, would represent a level of craftsmanship above the usual motor projects one tends to see. I wanted motors that would run well, if not smoothly. I wanted motors that would make creative use of the kinds of parts and materials that everyone has access to, including household cast-offs, second-hand-store treasures, and garage sale and flea market finds. I wanted motors that were built with the kinds of hand tools that any average person would have or could get.

I also wanted motors that had character and aesthetic appeal. I can accept that the look of a commercial electric window motor, buried down inside of a car door, is of little consequence to anyone. But my motors, like most of the projects for which I am known, were intended to amuse, entertain, and inspire. I've learned that a beautiful machine will attract the attention of even non-technical people who, in handling, poking, and prodding it, will find themselves immersed in subject matter they'd have never considered before.

It took a year of contemplation, scrounging, and building, but when I was finished, I had constructed a set of five "marvelous magnetic machines." This is a book about those machines. It's not a cookbook, and there are no plans, at least not in the traditional sense. You can appreciate that there is little point in publishing a parts list for a motor made from junk, when the scrap in my garbage can is certainly different than the scrap in yours. On the other hand, I think that this book offers something of greater value, namely, insight.

To understand how something is put together, it is far less important to understand the parts than it is to grasp the relationship between the parts. If you understand the relationship between parts, you're no longer tied to collecting specific items on a parts list. You're set free to make creative substitutions and to make use of sources for parts and materials that would not have been evident to you otherwise.

In the pages to follow, you will find ample details of my machines, but what I really hope to share is my thought process, my flashes of inspiration, and yes, even my mistakes. If, in reading this book, you are left with the impression that you had just sat beside me at my workbench and watched me as I built them, then I will have achieved my goal.

You have not read a book like this before.

Symbols Used for Units of Measurement

This book employs the symbols recommended by the International System of Units (SI) for expressing units of measurement.

The first time a unit of measurement is used in the book its name will be written out with its symbol following it — e.g. ampere (A). Thereafter only its SI symbol is used.

The SI symbols used in this book are:

ampere	A	second	s
hertz	Hz	volt	V
millisecond	ms	watt	W
ohm	Ω		

Safety

Imagine standing at a gas station fueling your car when, suddenly, alongside of you, a very expensive Italian-made sports car appears. You recognize it as a rare vehicle, one that can't be purchased for less than a quarter-million dollars. The owner opens his driver door, and bangs it hard against the gas pump, two or three times, as he gets out. He sets up the pump to deliver the cheapest gas the pump sells, then walks into the station to buy some chips and a soda. You can see through the windshield that he's left a lit cigarette on the dashboard, and it's now burning its way into the leather.

Next, he comes out with his chips, and noticing the pump hasn't finished yet, tosses his keys on the hood as he sits on one of the fenders. The keys have scratched the paint, and his rear-end leaves a dent in the fender. When the pumping has finished, the owner gets into his car, starts the engine, and stomps on the gas. The engine screams all the way down the highway because, though the car is moving at full speed, the driver hasn't bothered to take the transmission out of first gear.

Can we agree that, at least for the moment, this is still a free country? Assuming the owner paid for that car, it's his, fair-and-square. He can do with it as he pleases, even if that means destroying it. Still, who would not regard this person as an absolute fool? Surely, the most basic care and a little bit of caution would go a long way to protecting a valuable asset like this. So why wouldn't he do that?

The human body represents one of the most exquisite pieces of machinery ever devised. The intricacies of the human eye, the complexity of the human ear, and the flexibility, dexterity, and precision of the human hand exceeds anything mankind has ever designed. These, along with the rest of our body parts, are irreplaceable. Their value lies beyond classification with a price tag.

Most of us get a functional set of pieces and parts when we are born, yet for some reason, we tend not to care for and protect ourselves as we should. We scoff at the idiot who is careless with his expensive sports car, yet our tendency is to treat our own priceless and irreplaceable assets even more carelessly. Don't be like the guy in my story. Don't be a fool.

Protect your eyes! Existing medical technology can do little to repair a tattered eye, and even less to replace it. Safety glasses provide cheap insurance, and are available in various models at hardware and home-improvement stores. Take the time to read and understand the literature that comes with protective eye wear. Make certain that you understand the capabilities of your glasses, and that they're appropriate for the work you're doing.

Protect your ears! The mechanism that enables you to hear is exceedingly delicate. One hundred million years of evolution could not have prepared it for the cacophony that makes up the modern sound-scape.

Obviously, extreme noises can result in an instant and permanent loss of hearing. What many fail to realize, however, is that long-term exposure to even moderate sound levels is just as harmful. Bear this mind as you mow the lawn, blast the stereo in your car, or visit a dance club, concert hall, or pub. Think about your ears whenever you swing a hammer, push a plank into a table saw, or drill a hole into a piece of metal.

Foam rubber ear plugs or approved headphones are critical pieces of safety equipment. When selecting hearing protection, look for a decibel value, printed somewhere on the packaging material, that indicates the amount of noise reduction provided. The larger the number, the better.

Protect your lungs! As a young man, I worked for several years at a medical company that leased oxygen equipment. We had two principle customers: ex-smokers, and ex-coal miners. Both had destroyed their lungs inhaling contaminants, and both spent their remaining time on Earth struggling for every miserable breath they took. It's not a fate I'd wish on anyone.

Drilling, sawing, sanding, and grinding operations produce dust. None of this material is very good for your lungs. Masks are cheap, particularly the disposable "surgeon" type. Buy them, and wear them. You'll be surprised at the stains that appear in the filter material, even when breathing supposedly "clean" air.

When using paints, lacquers or other volatile compounds, make sure to work in a well-ventilated area. Lots of solvents are downright corrosive to lung tissue, and ordinary masks won't protect you. The more dilute you keep those fumes, the better off you are. As much as possible, I try to handle those sorts of materials outdoors, or in my garage, with the door wide open.

Protect your hands! Any process that can shape, cut, or drill wood and metal will do far worse to flesh and bone. Be cautious, and exercise good judgment in handling tools and materials. When using knives, cut away from your body, not toward yourself. If you're pushing small stock through a table saw, use a piece of scrap wood as a "pusher," not your fingers. Don't hold small parts in your hands when drilling. If the drill bit seizes in the hole, the work will whirl around unexpectedly, causing deep and painful cuts. Avoid wearing jewelry when operating power tools, and avoid loose sleeves or other clothing that could get wound up in rotating machinery. Wear gloves when handling chemicals.

Be alert for fire! Some of the projects that follow involve materials and techniques that, if mismanaged, could result in an accidental fire. Exercise good judgment when using propane torches or similar heat sources. Try to work in a clean area, free of combustible debris. Make sure that soldering irons are unplugged, cool, and properly stowed when you leave your work area. Use care in handling batteries to ensure that they are never accidentally short-circuited. Be conscious of the fact that many paints and solvents are extremely flammable. They can be set ablaze with nothing more than an electric spark. Protect your work area with an approved smoke detector (they are cheap!) and a suitable fire extinguisher. Know how to use it.

Know what you are working with! In years past, industrial workers often handled potentially hazardous materials in complete ignorance. Fortunately, regulations now mandate that each substance used in the workplace be documented with a so-called "Safety Data Sheet". Safety Data Sheets, also known as "Material Safety Data Sheets", describe the physical attributes of various chemicals, as well as crucial information pertaining to storage, handling, disposal, and possible health issues.

Hundreds of thousands of Safety Data Sheets are available for free to anyone with access to the Internet. Visit your favorite search engine and conduct a search on the phrase: "SDS", "Safety Data Sheet", or "Material Safety Data Sheet".

Just for fun, take a look under your kitchen sink and retrieve some laundry bleach or drain cleaner. Write down the chemicals that appear in the ingredients, and then look them up. The exercise is very instructive, if not alarming.

Pay attention! When you're working with tools or conducting experiments, needless diversions or distractions will lead to mistakes. Mistakes may translate into injury. Try to work someplace quiet where you won't be interrupted. Unplug the television. Most people watch too much anyway. Don't drink, smoke, or eat anything that's likely to diminish your mental faculties. The most powerful safety device in existence is a clear head and a mind focused on the task at hand.

Don't be afraid to ask for help! I strongly advise younger experimenters to enlist the aid of a mentor. A parent, a teacher, or other responsible adult can provide insight, guidance, and even training. Their wisdom can prevent accidents.

When I think about my own youth and some of the "experiments" I attempted in years past, it makes me laugh, albeit nervously. I don't remember being reckless, it's just that incomplete knowledge and simple inexperience can place a young person at risk. The fact that I reached adulthood unscathed is as powerful an argument for guardian angels as you'll ever find.

While my father was not scientifically inclined, he was an old-school craftsman of the highest order. He went to great lengths to teach me the proper use of tools and some degree of common sense. That alone has probably kept my "bacon out of the fire," and for that I'm eternally grateful.

The bottom line: In this age of blame-shifting and litigation, it's tempting to try to anticipate every possible mistake a reader might make and then issue the appropriate warning. That's a game that no author can win.

Please, use your own common sense. Ultimately, ***you*** are responsible for what you do, and for your own safety. If something an author suggests does not seem safe to you, ***don't do it***. If a process or technique requires more skill than you possess, ***don't do it***.

If you are one of those individuals that God blessed with two left hands, please be content to enjoy books like this one for their academic value, and leave well enough alone.

Chapter 1
Principles

There is a certain logic to the way I set out to build my machines. If you're not familiar with the principles on which that logic is based, you may find yourself wondering why I did what I did. For this reason, it's worth it to take a few moments to muse on the subject of magnetism itself. Of course, this is not the book for an in-depth study of the topic, but that doesn't mean I can't draw your attention to a factual nugget or two that I think will come in handy later.

The Earth as a Magnet

Most humans regard the earth as a large, rocky sphere. In truth, only the very surface, to a depth ranging from a few miles to a few tens of miles, it is composed of rock in the sense we normally think of it. Beneath the crust is a layer called the mantle, and below that is the core of the earth. The core itself is composed of at least two layers. The inner core is a solid, metallic sphere at the very center of the earth, suspended in an outer core comprised of molten metal, a relationship that reminds me of a cherry cordial. The core is relevant to our larger discussion, because a so-called "dynamo theory" postulates that motion in the liquid outer core (or possibly a liquid sub-core within the solid inner core) is responsible for the creation and long-term maintenance of the earth's magnetic field. This field emerges near, but not precisely at, the southern axis of the globe, extends into space, curves around, and reenters the earth in the vicinity of, but not precisely at, the northern axis.

This field is sufficiently intense to leave its permanent imprint on certain metallic rocks and minerals. Rocks containing hematite and magnetite, for example, can be magnetized by exposure to the earth's magnetic field. Samples of these materials can be brought into the lab and analyzed, revealing the history of and changes to the earth's magnetic field over many millions of years. At some point in our distant past, people noticed these magnetic rocks, sometimes referred to as lodestones, and their peculiar properties. People noticed that lodestones would sometimes stick together, though sometimes they repelled each other. When iron came into use somebody surely noticed that bits of iron would cling to lodestones as well.

The first practical use of magnetism was in the construction of an instrument synonymous with navigation: the compass. If a piece lodestone is suspended from a thread, or placed upon a slab of cork floating in a bowl of water, the lodestone magically orients itself so as to always come to rest "pointing" in a specific direction. Of course, what is really happening is that the lodestone is reacting to the earth's magnetic field, and aligning itself with respect to that that field.

When I was in school, we were taught that the compass was invented by the ancient but technologically-advanced civilization of the Chinese. Depending upon what you're willing to accept as evidence, this certainly occurred before 1000 A.D., and maybe as far back as 400 B.C. The discovery of a crafted piece of magnetized hematite, shaped into a rod-like form, complete with a groove that could be used for sighting the instrument, has been found in connection with the Olmecs of Mezoamerica. If correctly interpreted, that artifact has the potential to push the invention of the compass all the way back to 1000 B.C.

Later, somebody discovered that it was possible to confer the properties of magnetism onto a previously unmagnetized iron needle by stroking the needle with a lodestone. This, of course, was useful in the construction of more delicate and sensitive compasses. It seems the Chinese are credited with this development as well, though I suspect that, given the magical properties of magnetism, and the experimentation that it always seems to inspire, knowledge of the properties of lodestones, magnetic materials, and compasses may have developed independently in numerous places over a wide span of time. Having watched my grandson in his experiments, and recalling my own, it would not surprise me if an archaeologist someday proves that much of this early research was accomplished in play by children.

Pairs of Poles

Magnetic poles always exist in pairs. Poles are the terminals from which magnetism flows. A classic way to demonstrate this is to tape a bar magnet to the underside of a sheet of white cardboard. If the top side of the cardboard is then sprinkled with iron filings, the particles of iron will arrange themselves according to the magnetic lines of flux as they exit one pole, curve about, and reenter the other pole. Note that magnetism is not actually composed of lines, though the effect of the magnetic field on the iron particles develops an image of flux lines that are useful for visualizing the operation of magnetic circuits. Television programs like to warn viewers, "Don't try this at home." On the contrary, this is one of those experiments that you really should try yourself, though a helpful suggestion is in order. Before you begin, seal your magnet inside of a plastic sandwich bag to protect it. If you should accidentally spill or scatter filings on a naked magnet, cleaning it can become an exercise in utter futility.

I mentioned earlier that the earth generates a fairly intense magnetic field with one pole lying at the far North (near, but not at the earth's axis) and the other in the far South. The fact that a bar magnet will align itself with the earth's field in a predictable way provides a ready means by which the poles of a magnet can be distinguished from one another. When the suspended magnet has finally come to rest, the pole pointed toward the geographic north is identified as the "north-seeking pole" while the other becomes the "south-seeking pole." "north-seeking pole" and "south-seeking pole" are wordy phrases that are usually shortened to "north pole" and "south pole," respectively. Often a simple "N" or "S" will suffice.

With two bar magnets, whose poles have been identified by the technique above, a few moments of play yields one of the most basic of magnetic principles, namely, that like poles repel, and unalike poles attract. Bring the north pole of one magnet to the north pole of another, they will repel. Bring the south pole of one magnet to the south pole of another, they too, will repel. Bring the north pole of one magnet to the south pole of another, and they will attract each other.

Now, if you were paying attention, something about the prior discussion of compasses should bother you. How is it that the north pole of a bar magnet, when suspended, will point toward (be attracted to) the earth's North Pole? Being alike, shouldn't the two north poles repel? The answer yes, they would repel, if they were in fact alike. But they're not alike; they can't be! Think about it; the north-seeking pole of the compass can only be attracted towards the geographic north of the earth if the magnetic pole at that end of the planet is, in fact, a south-seeking magnetic pole! That is the case!

Conversely, the magnetic pole that lies in Earth's antarctic region is actually a north-seeking pole. By convention, lines of magnetic force (which we can visualize using the iron-filings-on-cardboard trick) are said to emerge from the north-seeking (north) pole of a magnet, and flow back into the the south-seeking (south) pole. That's why I said earlier that the magnetic field of the earth emerges at the magnetic pole near the geographic south and flows to the pole near the northern axis.

Force and Distance

At this point, it probably goes without saying that the attractive and repulsive properties of magnets can be exploited to exert force on moving parts of a machine, and are therefore the basis for the construction of electric motors.

If we had a time machine, we might chose to visit the laboratory of Charles Augustin de Coulomb (1736-1806). Coulomb is best known for his contributions to the understanding of electricity and electrostatic forces, but he worked with magnetism as well. In a series of experiments in which the force between two magnets was measured by a specially-designed balance, Coulomb derived a formula which expresses the force between two magnetic poles, given the strength of the magnets and the distance between those poles. That expression can be found in figure 1-1.

F is the measure of force between the magnetic poles. Q1 is the strength of one of the magnetic poles, Q2 is the strength of the other pole. K is a constant multiplier, and d specifies the distance between the poles.

$$F = \frac{kQ_1Q_2}{d^2}$$

***Figure 1-1**: Formula for attractive force between magnetic poles.*

Just by looking at the relationship between these variables, without actually grinding any numbers, we can see that the force is directly proportional to the strengths of the magnetic poles involved. The higher the strength at the magnets' poles (either Q1 or Q2), the greater the force produced.

However, that force is inversely proportional to the square of the distance. What this means in plain English is that the farther apart the poles are, the weaker the force is between them. The force doesn't weaken proportionately to increases in distance, rather, it weakens as a square of the distance. If you double the distance between poles, the force decreases to one-fourth. Quadruple the distance, and the force decreases to one-sixteenth. Increase the distance by a factor of ten, and the force will weaken by a factor of one-hundred.

The point to walk away with is that magnetic force weakens rapidly with distance. If we want to exploit magnetism to make a motor turn, we can maximize the force produced by minimizing the gap between interacting magnetic components.

Permeability

The magnetic flux issuing from the poles of a permanent magnet has the capacity to induce magnetism in nearby materials, assuming those materials have the correct properties. We have already seen this effect in the alignment of iron filings as they arrange themselves in the field of a bar magnet.

Even a child knows that certain materials like iron or steel will stick to a magnet, while other materials, like wood or copper, seem indifferent to the presence of the magnet's field. These differences can be quantified in a number that expresses the material's propensity to carry magnetic flux inside of itself. That attribute is called permeability, and its magnitude is usually represented by the Greek letter mu (μ).

Even the vacuum of deep space has a permeability value associated with it, defined as $4*\pi*10^{-7}$ Henries/meter. This value is referred to by the Greek letter μ with a subscript of zero, or mu-naught. By itself, it tells a motor builder little, unless we use it as a yardstick to compare against other materials. To do this, we simply divide mu-naught (μ_0) into the actual permeability values of the other materials we might be interested it. The resulting value is called the relative permeability. Since μ_0 divided into itself is simply one, the relative permeability of free space is also defined as one.

The μ value of a magnetic material like cobalt, divided by mu-naught, yields a relative μ on the order of 250. This means that upon immersion in a magnetic field, cobalt will support an internal flux density that is 250 times greater than that which would be found in free space. The relative μ of pure nickel is around 600. Pure iron, annealed in hydrogen, has a relative μ of about 200,000. Supermalloy, an alloy composed of a mixture of nickel, molybdenum, and iron, can exhibit a relative μ on the order of one million. In general, high permeability means strong "ferromagnetic" behavior.

Actually, all materials, including the apparently nonmagnetic materials we cited earlier, exhibit some kind of magnetic behavior. However, for practical purposes, materials with permeability values near one can be regarded, for our purposes, as being non-magnetic.

Interestingly, in some materials, the relative permeability is actually less than one (less than free space) The effect of this is that the material is repelled by the poles of a magnet, regardless of whether the pole is north or south-seeking.

This behavior, termed diamagnetism, is very weak, and not usually observed in day-to-day life. However, in the case of pyrolytic graphite, which is both lightweight and diamagnetic, exposure to strong magnets yields repulsive forces sufficient to cause it to levitate. At present, the effect is of little practical use, but is very interesting to observe. Examples of this behavior can be seen on Internet video sites like YouTube. Go visit YouTube or a similar video site and conduct a search using the keywords "pyrolytic graphite levitation."

Magnets on Demand

Another great stop on any time machine itinerary would be the lecture hall in which Hans Christian Oersted (1777-1851) gave a presentation in the spring of 1821. Oersted was aware that ships' compasses were sometimes affected by lightning strikes, and he probably knew that Benjamin Franklin had succeeded in magnetizing steel needles with electric discharges from Leyden jars. Oersted had suspected a connection between electricity and magnetism, but it was during one of his lectures that he laid the groundwork for definitive proof. He passed an electric current from a battery through a thin platinum wire, and noticed that the needle of a nearby compass moved by itself. A compass needle, which at rest lay parallel to the de-energized wire, suddenly rotated to a perpendicular position when the electricity was turned on. It took several months of additional experimentation to fully qualify and understand what he had seen, but his findings were destined to change the world.

We have already mentioned the north and south poles of magnets and that, according to convention, the lines of magnetic flux flow from the north pole to the south. Iron filings can give us a visual picture of those lines. When we draw them, we sometimes add little arrowheads to denote the direction of the flow.

Oersted discovered that when electricity passes through a wire, a magnetic field is produced, and the lines of flux that result circulate around the wire in concentric circular loops. The direction of this circulation depends upon which way the electrical current is flowing.

Over the years, people have contrived memory aids to assist in remembering the relationship between the flow of electricity and the direction of the magnetic field around a wire. One such aid is the so-called "right-hand rule." Imagine grabbing a wire with the right hand, wrapping the fingers around the wire, with the thumb extended in the direction of conventional (plus to minus) current flow. Do so, and the fingers wrapped around the wire point the way in which the magnetic field circulates. See figure 1-2.

Imagine, for a moment, a straight piece of copper wire rising off the face of this page, toward your face. Furthermore, imagine that the end of the wire closest to the page is connected to the positive terminal of a battery and the end of the wire closest to your face is connected to the negative terminal. In this case, conventional current flows from the page, up the wire, toward you. Wrap your right hand around that wire with your thumb pointed towards you (this is the direction of conventional electric current flow) and your fingers will tell you that the field whirls around the wire in a counter-clockwise fashion.

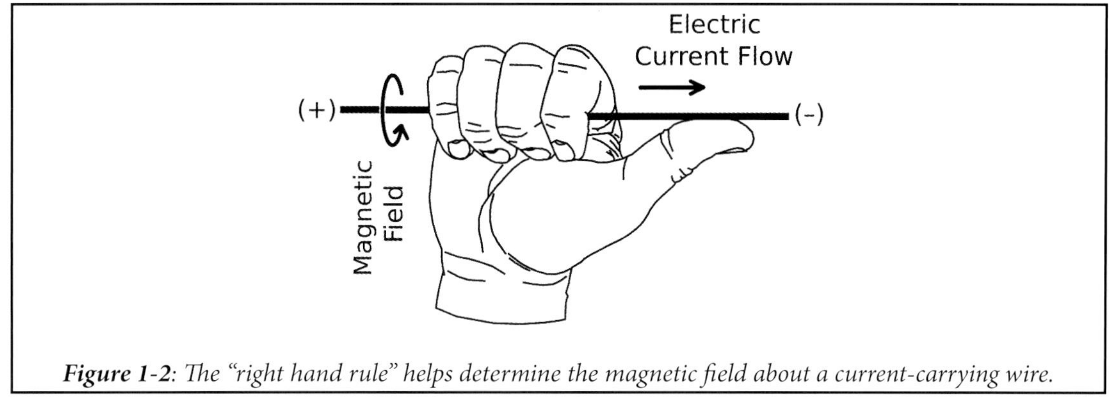

Figure 1-2: The "right hand rule" helps determine the magnetic field about a current-carrying wire.

If you place a compass next to the energized wire, you will notice, as Oersted did, that the compass needle aligns itself with the flow of the circulating field around the wire. If you reverse the battery connections, so that the current flows from you towards the page, the field around the wire will change direction, and the compass needle will flip to point the other way.

This demonstrates that with an electric current, we can not only produce a magnetic field on demand, we can control its polarity by changing the direction in which we apply the electric current.

André-Marie Ampère (1775-1836), for whom the electrical unit ampere or A is named, was able to define a relationship between the strength of an electric current in a wire, and the strength of the magnetic field produced around it. The field, in fact, is proportional to the current in the wire. Increase the current, and you increase the field strength. Yet, even with moderately high currents, the resulting magnetic field can do little more than bump a compass needle.

However, if we bend the previously straight current-carrying wire into a circular hoop, the lines of magnetic flux that whirl about the wire combine in a cooperative fashion at the interior of the loop. This cooperation results in a composite field of much greater strength.

If we then lengthen the wire so as to allow for the winding of a coil composed of many loops, we benefit from the combined field produced by each of those loops. The resulting electrically-powered magnet, or electromagnet, can be very powerful. The formula in figure 1-3 describes the flux, B, produced by a coil of N loops or turns, carrying L ampere of current, with a length of l (assumed to be long compared to the diameter of the coil). We've already discussed μ_0, so its definition should be evident.

$$B = \mu_0 \frac{NL}{l}$$

***Figure 1-3**: Formula for calculating magnetic field strength.*

The addition of loops to the electromagnet increases the magnetic field produced in the same way that increasing the current in the wire does. So, a coil with ten turns of wire carrying one amp of current produces the same magnetic field as a single-turn coil carrying ten amps of current. In fact, engineers sometimes roll the two parameters into one, called NI or ampere-turns. A coil which requires ten ampere-turns to produce the desired flux "B" can, in theory, be implemented with either of the two numerical options just mentioned. It could also be implemented as two amps into five turns of wire, five amps into two turns, or any other pair of numbers "N" and "I" whose product equals ten.

If we wind a coil with hundreds or perhaps thousands of turns of wire, the ampere-turn product is very high, even for low values of electrical current, and the magnetic field produced will be proportionately strong.

The relative permeability of free space is pretty small when compared with the μ values of ferromagnetic materials like iron. It stands to reason, then, that if we fill the interior of the coil with something other than free space, the larger μ values will lead to a much larger flux density in the core of the electromagnet. A cylinder of iron, for example, sized to fit inside the coil, is an ideal candidate. This final form of electromagnet, comprised of many turns of wire wound around a ferromagnetic core, is called a solenoid.

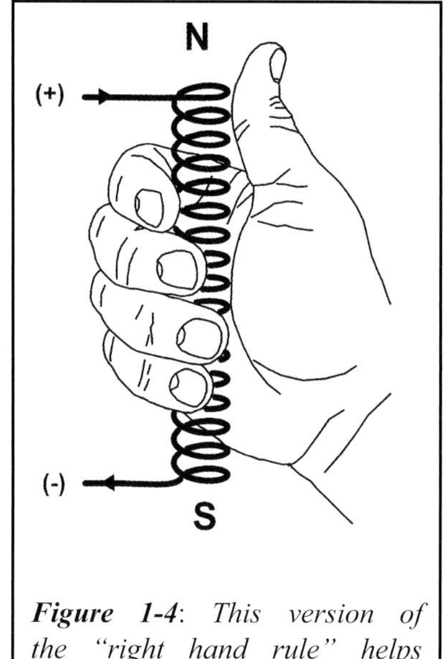

Figure 1-4: *This version of the "right hand rule" helps determine the field in a solenoid.*

To determine the polarity of the field emanating from the ends of the coil, we can invoke another incarnation of the "right-hand rule." In this, we grasp the coil with the right hand so that the fingers wrap around the coil in the direction of conventional (plus to minus) current flow. If the thumb is extended, as a hitchhiker might, it will point toward the north-seeking pole of the electromagnet. See figure 1-4.

Rusted-Out Vans and Permanent Magnets

Quite a few years ago, some friends of mine and I took a stab at a career in music. We wrote our own material, and we played with some regularity in clubs and bars in and around Detroit. This necessitated the purchase of a vehicle to haul all of our gear around. Money was tight, so my first van was a rusted and well-worn shell running on its second transmission and its third engine. When it was running, the engine clattered like a sewing machine, and the rear end roared and sometimes belched clouds of blue smoke. It was an infernal machine.

The steering box, the gizmo that translates the motion of the steering wheel into the angular position of the front tires, was hopelessly worn out. If you drove down a straight length of road with the steering wheel centered, you'd eventually notice that the van would wander to the left. If you slowly rotated the steering wheel to the right, in order to compensate, you would notice that nothing would happen until the steering wheel had advanced a fair bit. Only then, would you get a sense that the tires were responding and starting to straighten out. If you went a little too far to the right, it was not sufficient merely to recenter the steering wheel. You'd have to adjust the wheel well to the left-of-center in order to get the front wheels to straighten again. The mechanic who eventually fixed the problem called this defect "slop" in the wheel. Another referred to it as "backlash."

For the sake of argument, let's take the trouble to actually graph the behavior of a normal steering system, and then compare that to a graph depicting the behavior of my van. We can see an example of this in figure 1-5. The horizontal "X" axis represents the position of the steering wheel. The right side of the graph represents clockwise motion of the steering wheel, the left side of the graph represents counter-clockwise motion. The vertical "Y" axis is a measure of how far the front tires have steered. The upper half of the "Y" axis represents the extent to which the tires have turned to the right, and the lower half represents steering movement to the left. The condition where the steering wheel and tires are centered, is the point represented by the intersection of the "X" and "Y" axis. Got it?

Now let's take a look at properly-functioning steering gear. Starting with the tires and wheel centered at point (a), we turn the steering wheel clockwise. The front tires, of course, move to the right. Turn the wheel more, and the tires move proportionately. If we continue to turn the wheel until it will turn no farther, and if we make a series of measurements and plot the data all along the way, we get the segment (a)-(b) on the graph in figure 1-5. It starts at the intersection of the "X" and "Y" axis (the zero point) and slopes gently upward as it extends to the right.

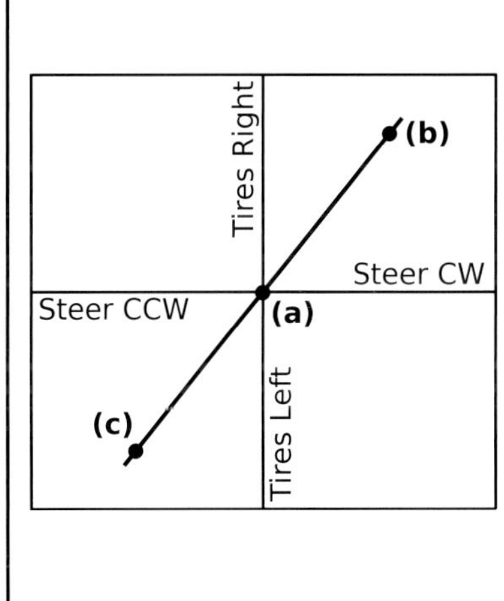

***Figure: 1-5**: This graph depicts a steering system that works properly.*

Now let's turn the steering wheel slowly counter-clockwise, as far as it will go, and continue to record and plot data throughout the process. The first thing we notice is that we get the same data moving left as we did moving right. You can replot that data, but you end up drawing right on top of segment (a)-(b).

At some point we reach the centered-wheel and centered-steering point. As we continue to move the steering wheel counter-clockwise, we develop the line segment (a)-(c), which starts at the intersection of the "X" and "Y" axis, and descends gently to the left. For good measure, when we've gone as far to the left as possible, we can finish things off by returning the wheel to the centered, neutral position. Now we review the data we've collected.

The plot of the mechanism's behavior is a straight line (c)-(a)-(b), which slopes gently from point (c) at the lower left, through (a) at the intersection of the "X" and "Y" axis, and rises gently to point (b) at the upper right. A straight line means proportionality; that is, we can conclude that the front tires respond directly to the position of the steering wheel. Furthermore, the slope of the line represents the "speed" of the steering box. The steeper the slope of the line, the more aggressively the front tires respond to changes in the steering wheel's position, and the more "touchy" the feel of the steering will be.

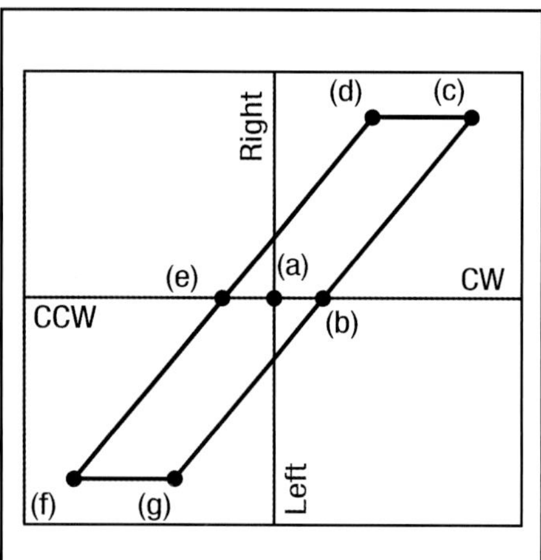

Figure: 1-6*: This steering system suffers significant backlash.*

Now, let's consider the steering of my old van by looking at the graph in figure 1-6.

Assuming the steering wheel is centered and the front tires are straight, we start at point (a). We turn the steering wheel clockwise, but for a while, there is no movement of the front tires. This is because of the "slop" in the steering box. This accounts for a horizontal line segment (a)-(b). Then, the steering gear begins to respond, and the front tires start to turn with the movement of the steering wheel. By the time the steering wheel has turned as far clockwise as it will go, we have plotted the line segment (b)-(c).

Now, we throw the steering wheel counter-clockwise. Again, because of the slop or backlash in the steering box, the front tires don't immediately respond. This yields line segment (c)-(d). At some point the front tires begin to steer to the left, and if we continue to rotate the steering wheel until it has gone as far counter-clockwise as it will go, we eventually plot the line segment (d)-(e)-(f).

Finally, if we rotate the steering wheel clockwise once more, we see more evidence of slop. We get no initial movement of the front tires, resulting in line segment (f)-(g), followed by movement recorded by the line segment (g)-(b).

What conclusions can we draw from comparing the first graph to the second?

In the first graph, the position of the vehicle's front tires is directly proportional to the position of the steering wheel. In fact, for any given position of the steering wheel, this translates to exactly one angle of displacement of the front tires.

In the second graph, there is no single line that defines the relationship between the steering wheel and tire position. Rather, the graphical relationship is a sort of open, slanted box-like affair. That means that we cannot infer with certainly the position of the front tires based on the position of the steering wheel. In fact, if we return the steering wheel to the centered position, the tires may still be skewed as far to the right as point (b), or as far to the left as point (e). The position of the front tires depends not only upon the position of the steering wheel, but whether the steering wheel was brought there from some prior position farther clockwise or farther counter-clockwise.

The technical term for this "slop" or "backlash" is "hysteresis".

An interesting implication of hysteresis in a defective steering mechanism is that it exhibits a kind of "memory". If I turn the steering wheel clockwise to make a right-hand turn, and then straighten out the wheel on the completion of that turn, the front tires are likely to remain skewed slightly to the right. It is not enough simply to straighten the wheel to correct this skew, because the hysteresis in the steering gear causes the front tires to retain their last position until I move far enough in the opposite direction to clear that condition.

Now what can all of this possibly have to do with magnetism? The answer begins with a fresh graph, figure 1-7.

First we rename the horizontal axis "H". "H" represents the strength of an applied magnetic field. The vertical axis we shall label "B." "B" is the magnetic flux density in a material that results from the immersion of that material in the magnetic field. If we generate a slowly increasing magnetic field in free space, measure the resulting magnetic flux as the field rises, and plot that information on the graph, we end up with a line that describes the relationship between "H" and "B." In free space, for example, the plot takes the form of a straight line that slopes gently upward from the lower left to the upper right. The steepness of that incline represents the propensity of free space to support magnetic flux within itself. This should sound familiar, because this is how we defined permeability a few pages back.

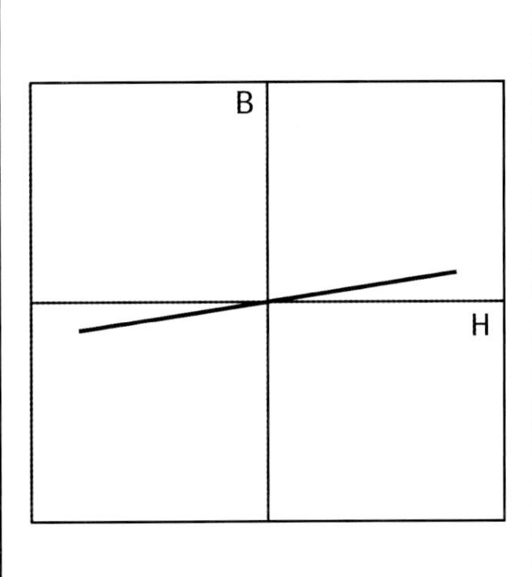

***Figure: 1-7**: This is a B-H plot representative of free space or a substance like air.*

In the case of the steering graph, the steeper the line, the more "responsive" the steering system is. In this case, the steeper the line associated with a given material, the higher its μ, and the more responsive that material is to immersion in a magnetic field.

Materials we commonly regard as magnetic, like iron or steel, have very high μ values, which suggests that their μ line plots should rise very steeply.

Well, they do... sort of. Iron and steel are classified within a group of substances called ferromagnetic materials. Ferromagnetic materials are interesting not only because of their large μ values, but because their μ values are not necessarily constant.

Figure 1-8 depicts the B vs. H plot for a hypothetical chunk of steel. Where there should be a straight line, we instead see an open S-shaped curve that looks a lot like the behavior of the steering gear in my old van. Ferromagnetic materials exhibit hysteresis. Let's take a tour of this hysteresis curve.

With no external magnetic field applied, there is no flux in the steel. The value of both "H" and "B" is zero, which means we begin at point "a," which is the intersection of the horizontal and vertical axis.

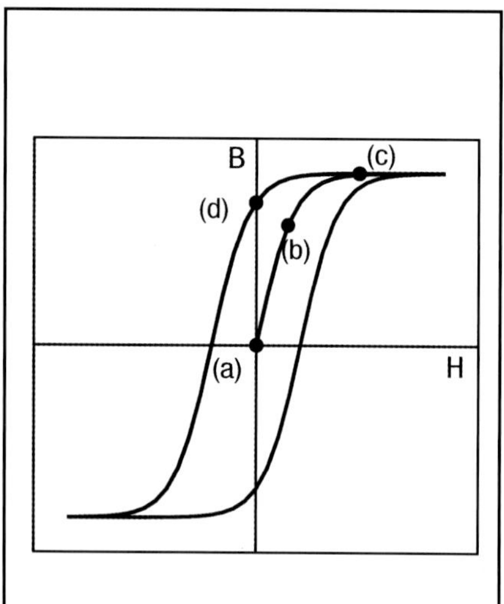

Figure: 1-8: The B-H plot for this magnetic material shows evidence of both hysteresis and saturation.

As we increase the applied magnetic field "H", we move along the hysteresis curve from point (a) to point (b) and upward. Notice that when we reach point (c), something interesting happens. No matter how much farther we increase "H", "B" simply flattens out and refuses to rise any higher. This is called the material's point of saturation. If this was the core of an electromagnet, no matter how many extra turns of wire we wound in the coil, and no matter how much extra current we passed through the coil, the flux in the core could not rise above the level represented by point (c).

If we now decrease "H", following the curve from point (c) to point (d), we can see that the "B" value does not go to zero even if "H" has. The hysteresis in the steel exhibits the same sort of memory effect that we observed in the worn-out steering system. In the steering example, memory manifested itself in tires that don't straighten out. In the steel, this memory is expressed as residual magnetism that is retained in the steel even after the external "H" field is removed. Instilling this kind of memory in the metal is more commonly referred to as "magnetizing" the steel.

Can we demagnetize the steel? We could attempt to drive the magnetic field in the opposite direction, so as to clear the prior condition, but one must be careful. Go too far, and you can leave the metal magnetized again, but this time in the opposite direction.

An energetic driver of the my old van might notice that if the steering wheel is whipped rapidly back and forth, from lock to lock, and this is done quickly and aggressively enough, the net result to the front tires is that they will center themselves and remain centered. The same idea can be used to demagnetize steel. If we wind a coil of wire around the steel, and apply an alternating current to the coil, the continual reversal of the electric current results in a magnetic field that is continually reversing. This thrashes the hysteresis curve in the steel the same way that the driver thrashes the steering wheel in the van. The net result is a "B" value of 0.

Hysteresis curves like that depicted in Figure 1-8 can be fat with a large open space in the middle. Fat curves represent materials that may require large magnetic fields to magnetize them, but which retain that magnetism when the external field is removed. The hysteresis may be so pronounced that demagnetization of the material is exceedingly difficult. These are sometimes termed "hard" magnetic materials.

On the other hand, hysteresis curves can be very skinny. They can be so skinny in fact, that the curve degenerates into a flattened S-shape with no open interior. The behavior of this type of material more closely resembles of the behavior of paramagnetic (straight-line) materials. As long as the external "H" field is present, the material supports a large amount of magnetic flux. The minute the external field is removed, however, the material ceases to be magnetic itself. It exhibits little of the memory we observed in fat hysteresis curves, so this type of material will not retain a magnetic field of its own. This could be considered a "soft" magnetic material.

The choice of hard or soft materials for use in the construction of a magnet depends upon the type of magnet we're building. In the case of a permanent magnet, like a compass needle or a refrigerator magnet, we want to use a material that, when exposed to a magnetic field, will exhibit memory of that exposure, and retain a permanent magnetic field of its own. The best material to use to manufacture such magnets would exhibit hysteresis curves that are fat and open, which means that we want to use a hard magnetic material.

If we are designing an electromagnet, we want a core with a high μ that will support the field produced by an applied electrical current, but will not retain that field when the electric current is removed. We want to select a core material with a thin, closed, hysteresis curve. We would therefore select a soft magnetic material for use in our electromagnet.

Forces on a Wire

Oersted demonstrated a connection between electricity and magnetism when he passed an electric current through a wire, and caused the needle of a magnetic compass to move. The flow of current creates a magnetic field around the wire, which then interacts with the needle of the compass.

Since the basis for the movement of the compass needle ultimately amounts to a magnet-to-magnet interaction, exactly which part moves really depends upon the relative mobility and mass of the parts that are interacting. The needle in the compass is light, delicately balanced, and easily influenced. But suppose it were more powerful, more massive, and fixed to the table? What if the wire was free to move? What then?

Figure 1-9 rearranges the components of Oersted's experiment to answer this question. A large, cylindrical, permanent magnet sits vertically on a table with its north-seeking pole facing upward. Two parallel, uninsulated, copper rods are positioned just above the face of the magnet, and are connected to a battery or similar power source.

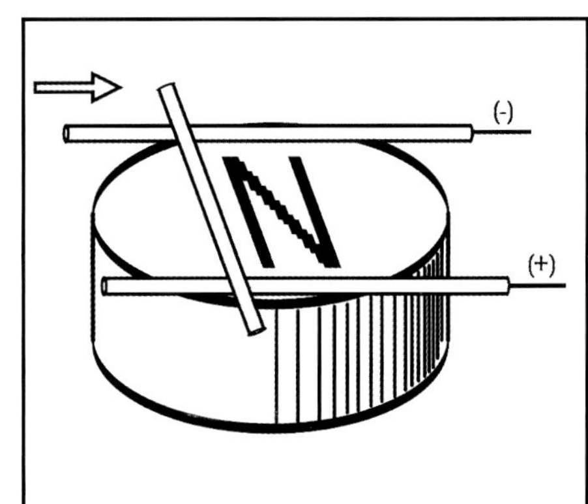

***Figure: 1-9**: The magnetic field produced by the current in the rod will cause it to roll to the right.*

To answer our earlier question, all we need to do is place a third copper rod across the first two. If the experiment is set up as depicted in the diagram, the rod will roll sharply to the right, seemingly under its own power. The force that produces this motion appears at right angles both to the direction of current flow and the direction of the permanent magnet's field.

We can gain some insight into why this might occur if we apply the right-hand rule to the current flow in the wire. In the diagram, current flows into the movable rod in a direction that is away from the observer. Thus, if we wrap our right hand around that rod, with our thumb pointing in the direction of current flow (again, away from the observer) we will see that the field produced around the rod circulates in a clockwise fashion. Note that this occurs within the field of the permanent magnet. Since we stipulated that the face of the magnet is north-seeking, flux can be thought of as streaming out of the magnet's pole face, in an upward direction.

To the left of the movable rod, the circulating field around the wire flows in the same direction as the field from the magnet. To the right of the rod, however, the circulating field of the wire is at odds with the field of the permanent magnet. Notice that the rod is repelled on the side where the fields are the same, and driven in a direction toward the side of the rod where the fields are partially canceled.

If we change the system so that the polarity of the permanent magnet is reversed, or we reverse the flow of electric current, and then repeat the experiment, the rod will roll to the left. If we reverse the polarity of the magnet and reverse the flow of the current, the rod will roll to the right once more.

Consider one more change to this basic idea. Instead of using copper rods, suppose we fashion a square loop of copper wire, as depicted in figure 1-10. The loop is suspended between the poles of two magnets and energized through slip rings as shown. Because of the polarity of the magnets, and direction of the current flow in the side of the loop closest to the bottom magnet, the force generated on the wire there will act to push it to the left. The current at the top of the loop is returning, and is therefore flowing in the opposite direction. The force induced on that section of the loop will be toward the right.

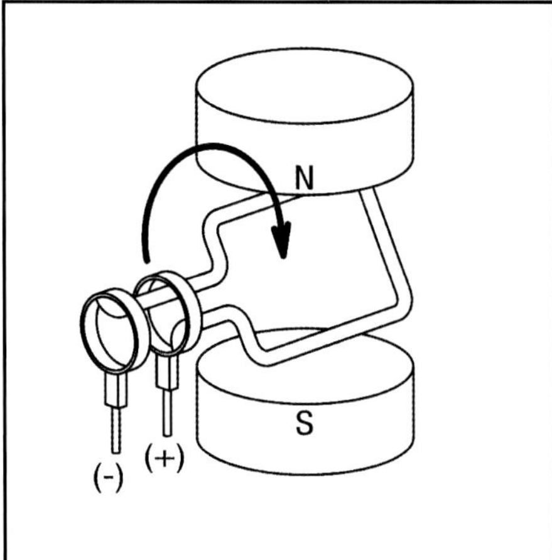

***Figure: 1-10**: With an electric current applied as shown, a torque is produced, inducing the loop to rotate clockwise, as indicated by the arrow.*

The combination of forces at the top and bottom of the loop imposes a clockwise twisting action, or torque, inducing the loop to rotate. Strengthened by the addition of multiple turns or loops of wire, and affixed to a shaft with proper bearings, it is readily apparent how this basic idea can be turned into a practical electric motor.

Summary

In this chapter, I introduced and covered a lot of material. Because this is not a physics book, my treatment of the material was intended to be general and introductory. This is likely to be useful to anyone with a casual interest in magnetism, particularly where it applies to building motors, but frankly, all of this subject matter is worthy of additional study.

Before we leave this chapter, let's summarize some key points:

- The earth is a big magnet. The ancients were able to exploit this for the purposes of navigation. Magnets have active surfaces called poles, whose polarity can be identified either as north-seeking or south-seeking. Poles always occur in pairs, so where there is a north-seeking pole on a magnet, somewhere, someplace, there must also be a south-seeking pole.

- Magnets can exert physical force on each other, through the interaction of their respective magnetic poles. Unalike poles attract each other, like poles repel. The amount of force exerted is proportional to the inverse of the square of the distance between them. The force drops off rapidly for even small increases in distance.

- All materials respond to magnetism in some fashion, but they differ in the extent to which they will support magnetic flux. This characteristic can be expressed in terms of the materials' permeability, or μ. Low-μ materials are generally regarded as non-magnetic. Steel and iron, materials that are regarded as being magnetic, have high-μ values.

- Magnets can be created on demand by passing an electric current through a coil of wire. The strength of the magnetic field produced when the current is turned on is proportional to the product of the current passing through the wire and the number of turns wound in the coil. The greater that product, the greater the magnet's strength. A coil with a single turn carrying 100A will produce the same field as a coil with 100 turns of wire carrying but one ampere.

- Some materials become magnetic if an external field is applied, but cease to be magnets when the external field is removed. Other materials can retain their magnetism even after the external field is removed, and are regarded as permanent magnets. This attribute is dependent upon the composition of the materials and a measurable behavior called hysteresis.

- Finally, if a current-carrying wire is placed in a strong magnetic field, the wire will experience a mechanical force which seeks to move that wire at right angles to the flow of current and the applied field.

Chapter II
The Peewee Motor

***Figure 2-1**: The Peewee Motor, front.*

No one less than the great chemist and microbiologist Louis Pasteur is said to have remarked, "*Chance favors the prepared mind*". Personal experience says that he is absolutely correct. My own predisposition to tinker with every piece of electronic or mechanical scrap I can get my hands on, combined with a chance event, led directly to the construction of the motor I am about to describe, and by extension, to the creation of this entire book.

I was at work, and I needed to print something. We have shared printers in our office, those large combination printer/copier/scanner affairs. I clicked my computer mouse to send off a print job, and then left my desk to visit the printer and pick up my paperwork. There, I ran into the fellow who maintains those machines. He had just finished repairing my printer and was about to toss some kind of part into the waste can.

"*May I have that?*" I asked. "*Why sure,*" he said, amused. Handing it to me, he asked, "*What are you going to do with it?*" "*I dunno,*" I answered honestly, "*but I'll think of something.*"

The details of that printer component would be irrelevant, if I could even remember them. The point is that, mounted upon it someplace, was a contraption I recognized as a solenoid. I removed it, and turned the part around in my fingers. This, I decided, would form the basis of a nifty little motor I subsequently named the "Peewee Motor". See figures 2-1 and 2-2.

Solenoids

In the last chapter, we discussed how to use an electric current to produce a magnetic field, namely, to pass it through a coil of wire. We also noted that if the coil is wound upon a suitable material, like iron or steel, the field produced is much stronger.

A useful variation on this idea is to wind the coil on a hollow tube. If we energize the coil, and then insert a short steel rod into the mouth of the tube, we will observe that the magnetic field in the interior of the coil will exert a force on the rod so as to draw or suck it into the interior. Turn off the juice, and the steel rod falls out. This arrangement of parts, often referred to as a solenoid, has the ability to convert an electrical current into a linear mechanical force.

Practical solenoids exhibit several enhancements that greatly improve their performance. Many solenoids feature an external steel frame that wraps around the exterior of the coil, which helps to further intensify the field produced by the coil. The bore of the coil may be plugged at one end with a section of steel rod, bonded to the frame. This is also intended to intensify the magnetism. This is sometimes referred to as a "stop". The moving part of the solenoid, referred to alternately as the "armature" or "plunger," is composed of materials with a high-μ. The end of the plunger may be machined to a cone-shaped point, and may engage with a similar conical cavity in the previously-mentioned bore plug. Again, this detail enhances the coil's effect on the movable plunger, so as to increase the force with which the plunger is drawn into the core.

***Figure 2-2**: The Peewee Motor, rear.*

Solenoids find use in endless applications. In the starter motor of your car, for example, a solenoid is used to engage the starter motor's pinion gear with the engine's toothed flywheel when you turn the ignition key. When you let off the key, the solenoid stops exerting its force, and the starter is disengaged. For many years, solenoids were also used to actuate electric door locks. Solenoids have been used to slap the steel ball around the inside of pinball machines, engage transmission components in washing machines, shift gears in video tape recorders, ring door chimes, lock hard drive read/write heads in their "park" position, and more. I've already alluded to the fact that they appear in modern photocopying and printing equipment.

A solenoid's behavior can be described in terms of several key characteristics. First, the quantity and gauge of wire that comprises the coil defines its normal operating voltage. Suffice it to say, for the moment, that one needs to be mindful of this specification so as not to overheat and destroy the coil.

Second, is the solenoid's actuating and holding force. Related to the number of ampere-turns in the coil, and the physical attributes of the frame and plunger, this represents the force exerted upon the plunger when the coil is on and the plunger is being drawn into the solenoid. Depending upon the size, design, and intended application of the device, this force may be measured in ounces or grams, or it could conceivably be measured in many hundreds of pounds or kilograms.

Finally, the plunger has a limited range of travel over which the solenoid will function properly. One limit of the travel is that point at which the plunger is fully drawn into the solenoid. The other limit is the point at which the plunger has been withdrawn so far that the coil can no longer exert a meaningful force upon it (actually, the outer limit of travel is also influenced by mechanical considerations, like proper alignment of the plunger with the bore of the coil, and consideration of the need to somehow retain the plunger so that it doesn't fall out completely when the coil is un-powered). The distance between these two limits, the span over which you can use the plunger to do useful work, is called the stroke.

Linear to Rotary Motion

Practical knowledge of electricity and magnetism first emerged during a period when steam power was already an established technology. It is not surprising, then, that numerous early inventors sought to apply steam engine principles to the construction of electric motors.

When steam is used to drive a piston, the resulting motion is linear. When combined with a connecting rod and a crank, that motion can be converted into the generally-more-useful rotary motion. Likewise, a solenoid plunger can be linked through a connecting rod to a crank.

Note that in single-acting steam engines, steam is applied to the face the piston, and the piston is pushed. The piston and connecting rod rotate the crank by pushing on it.

The Peewee Motor differs in that it is an attraction motor. Force is exerted by pulling on the crank, which occurs only when the solenoid is energized and the plunger is being drawn into the coil.

The operation of a steam engine requires precise timing. Steam pressure is applied to the piston, or relieved and vented to the atmosphere, at the command of sliding valve, like a spool or "dee" valve. The valve is connected through a linkage to an eccentric or cam. The eccentric and piston crank are both mounted on a common shaft. As the eccentric wobbles, it pulls and pushes on the valve actuator, which directs steam in and out of the engine's cylinder. The angular relationship between the valve eccentric and the position of the crank, otherwise referred to as the engine "timing," is critical. After all, we only want the piston to apply force to the crank when it is beneficial. At other times, we want the piston to move freely.

Correspondingly, the operation of the Peewee Motor depends upon precise timing. In this case, we aren't applying steam but electric current to a coil that acts on our "piston." The Peewee makes use of an electric switch that is turned on and off according to the position of the crank. I'll describe the specific details of that in just a moment.

With its single solenoid, the Peewee Motor is similar to a small single-action piston steam engine in another regard. If we consider the full rotation of the crank, neither the steam engine nor the Peewee Motor apply force for more than a fraction of the total rotation of the crank. This means that something must be done to help carry the crank through its full rotation so that the piston or solenoid can act upon the crank again. This is accomplished by adding a flywheel to the crankshaft.

Let us now consider how the office copier solenoid and a handful of other cast-off parts were cobbled together to create a working motor.

The Base

Tradition has it that you build little models like this on a slab of wood. Often, that means a chunk hacked off of the end of a plank. Some people invest little effort here, because they figure that making something nicer is too difficult. Not really. If you invest a little creative thought into the project, and maybe "think outside the box," you'll find loads of objects that, while never intended to be part of a model motor, can with some modification, be made to look as though they were custom-made. The bonus, here, is that most of the difficult steps of fabrication have already been done for you.

For example, if you're committed to using wood for your base, check garage sales, flea markets, and art stores for decorative plaques. These might have photos or artwork decoupaged on them, or cute or inspiring messages on their faces, but all of this can be eliminated with sandpaper or a coat of paint. Since these plaques often have nice beveled or routed edges, making use of these gets you the appearance of precision and craftsman-like finish without having had to do the actual work.

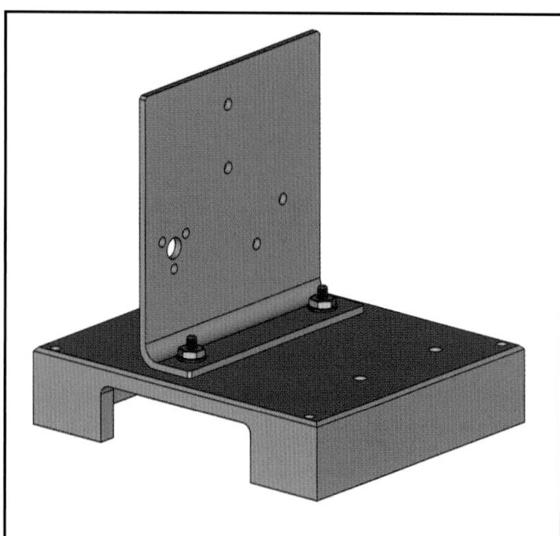

Figure 2-3: The Peewee Motor's base features an L-bracket to attach the machinery to the base.

Another option for motor bases are kitchen accouterments like shallow metallic dishes (inverted) or kitchen cutting boards. The bases for desktop decorations like pen and pencil holders are often made of metal or dense wood, are nicely finished, and would work well in a model motor application. Cast metal ashtrays, upright or inverted, offer some opportunity, as well.

In the case of the Peewee Motor, I looked at all of these options, but settled on a nicely machined piece of aluminum measuring 4 by 4-½-inches. The metal was junk, salvaged from who-knows what.

Attached to the base is an L-shaped bracket, also of aluminum. If I remember correctly, this was actually the heat-sink from a discarded power supply, but any comparably-sized and shaped part would have worked. The vertical portion of the bracket measures 3-½ by 3-½-inches. The foot of the bracket is about ¾-inch wide, and is anchored to the base with two 6-32 machine screws and nuts. See figure 2-3.

Bearings and the Flywheel

***Figure 2-4**: Flywheel, shaft, and bearing detail. Washers are added to the shaft, as necessary, to maintain spacing of moving parts.*

Experimenters who don't have skills or tooling to machine precision components are likely to be discouraged from building projects that require such parts. The trick is to be creative, and to imagine what second-hand objects might already contain the parts you need, ready for harvesting.

At one point, someone had given me a box of electronic junk which included several broken cassette tape recorders. Never content simply to throw things away, I dismantled the decks, salvaging transistors, audio connectors, transformers, springs and screws, among other things.

As you may well imagine, a motor is used to rotate the tape reels in the cassette, although this motion is not what actually advances the tape in the player. Advancing the tape across the tape heads simply by using the motion of the supply and take-up reels would result in very poor audio quality, because it's impossible to assure that the rate at which the tape moves will be smooth, regular, and uniform.

So, the reels are set up to apply no more force than is necessary to assure that the tape feeds and winds up neatly. The tape is actually advanced by threading it between a polished metal shaft, called a capstan, and a rubber wheel, called a pinch roller. Smooth capstan motion is guaranteed by having it run in a quality bearing, and fitting the capstan with a flywheel to dampen out irregular motion.

When I scrapped the cassette decks I just described, I took the trouble to remove the capstans and their attached flywheels, as well as the bearings in which they turned. These were the parts that made their way into the Peewee Motor.

The cassette bearing I used looks a lot like a little top hat. The "brim" of the hat, if you will, is a mounting flange by which the bearing can be attached to a surface. It was a simple matter to position the flange on the L-bracket, and drill a few holes for the mounting screws. I used 4-40 machine screws and nuts. While the bearing's body is made of cheap "pot" metal (probably some kind of zinc alloy), the interior is fitted with a bronze sleeve. Bronze is a common bearing material that wears well, while minimizing friction. A polished steel shaft will turn inside a properly-sized bronze bearing smoothly with very little drag. See figure 2-4.

With the bearing in place, I inserted the capstan shaft with its attached flywheel into the bearing and gave it a spin with my finger. It spun like a little top, just as it should. As the capstan shaft is no longer being used as part of a tape mechanism, and instead functions as the shaft on which the Peewee's crank is attached, I'll henceforth refer to it as the crankshaft.

The Crank

The heart of any reciprocating motor or engine is its crank. This is the Z-shaped contrivance that is responsible for converting the back-and-forth motion of a piston, or in this case, solenoid plunger, into rotary motion. It's a deceptively simple part to make, though there exists an intimate relationship between the dimensions of the crank and the attributes of the solenoid that will drive it. I'll address that detail in a minute, but for the moment, let me talk about fabrication.

A shaft collar, in machine terms, is a metal doughnut with a second hole, perpendicular to the first, that's been drilled through the side of the doughnut into its center.

The latter hole is tapped with threads, and fitted with a screw. If one slides the collar onto a shaft, and then tightens the screw, the collar is locked onto the shaft, and won't fall off. Loosen the screw, and the collar can be removed.

The crank in the Peewee Motor is comprised of two such collars, soldered to opposite ends (and opposite sides) of a short length of brass strip. A drawing of the crank can be seen in figure 2-5.

The collars themselves are easy enough make. You can cut short lengths of steel or brass rod to yield a couple of squat, little cylinders. Drill a hole through each cylinder from top to bottom. This hole is sized to accept the shaft that you intend to put the collar on. A second hole is then drilled perpendicular to the first, through the side of the cylinder, into its bore. Tap that hole, fit it with a set screw, and you're done.

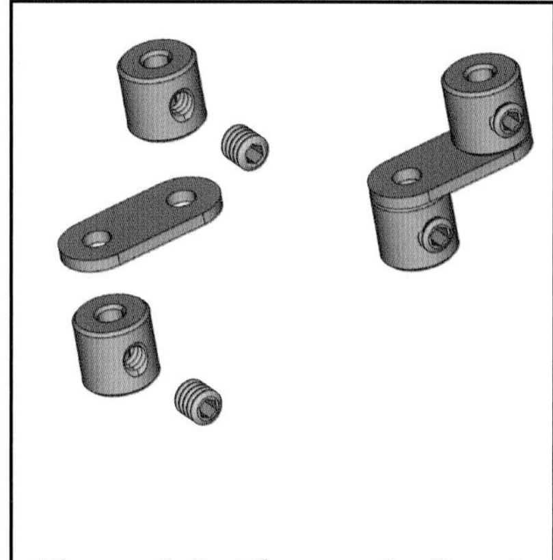

Figure 2-5*: The crank, first in pieces, and then soldered together with set screws in place.*

On the other hand, it's far easier to simply purchase collars. Any reasonably well-stocked hobby store, of the kind that caters to radio-controlled aircraft hobbyists, will stock them. They're used to hold model aircraft landing wheels onto their shafts, or to terminate the ends of the push rods that drive the aircraft's flight surfaces. A common brand is "Du Bro Dura-Collars." These are brass, plated to prevent corrosion, and come with Allen-type hex-drive set screws. Dura-Collars are available in different sizes to accommodate a wide variety of shafts, and each package even comes with a matching Allen wrench.

I cut my crank arm from 0.031-inch (1⁄32-inch) brass strip. The strip was cleaned with a scouring pad. I used a metal file to remove the plating and expose fresh metal on one face of each of my collars. I positioned the collars and clamped them in place with alligator clips. Then, I played the flame of a propane torch over the assembly and ran a bead of solder around the seams.

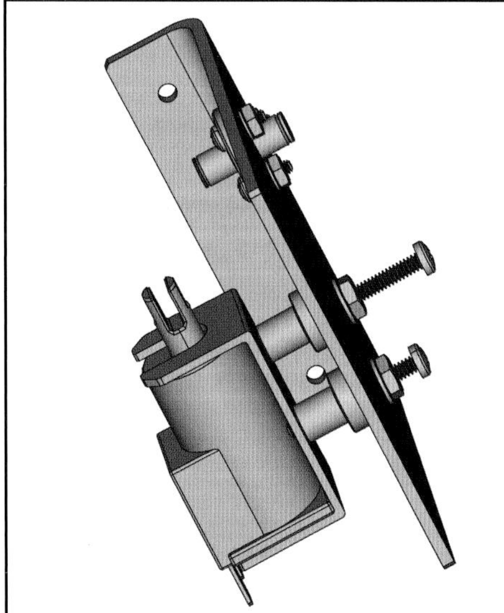

Figure 2-6*: Solenoid mounting details. The mounting screws are longer than necessary; washers, spacers, and jam nuts control penetration into the solenoid.*

The finished crank was installed on the crankshaft/flywheel assembly, and the set screw was tightened.

The next step was to install the solenoid and link the plunger to the crank with a connecting rod. Mounting details depend upon the specific make and model of solenoid, but the solenoid used in the Peewee Motor featured two 6-32 threaded holes in its frame. I drilled a couple of holes in the L-bracket to match. I inserted long screws through the holes in the bracket, added some spacers, and then threaded the screws into the solenoid's frame. This detail can be seen in figure 2-6. Jam nuts limit the penetration of the screws into the solenoid. The axis of the plunger was aligned to intersect the axis of the crankshaft.

The connecting rod is more properly termed a connecting link. It's nothing more than a short length of 0.031-inch ($^1/_{32}$-inch) brass strip, $^1/_4$-inch wide, and roughly 1-inch in length. I drilled two small holes in the link, one near each end.

The link couples to the crank end with a pin made from a piece of $^1/_{16}$-inch brass rod. A section of escutcheon pin or segment cut from a brad or finishing nail would do, as well. The pin is inserted and secured in the collar on the free end of the crank. The link is set on the pin, and the pin is terminated with another collar to keep the link from falling off.

The plunger of the solenoid I used was originally designed to connect to a linkage very similar to the one I fabricated for the Peewee Motor. The end of the plunger is slotted to accept the insertion of a link, and is drilled and tapped transversely to accept a screw that provides coupling and a pivot point between the plunger and the connecting link. Figure 2-7 shows how the connecting link ties the crank and solenoid plunger together.

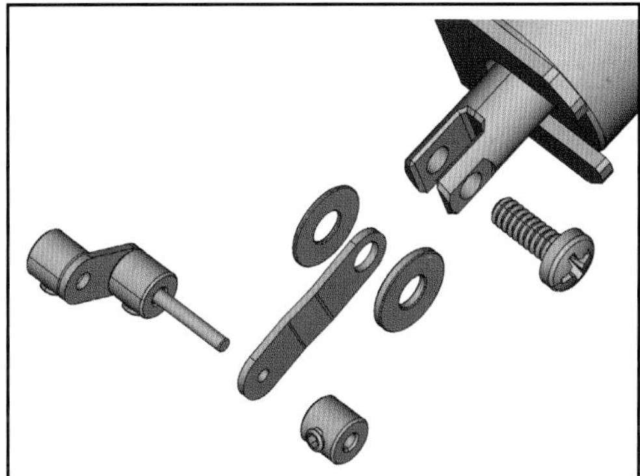

***Figure 2-7**: A connecting link joins the crank with the solenoid plunger. Washers are used to reduce play where the link attaches to the solenoid plunger.*

At this stage, I could spin the flywheel with my fingers, the crank would turn, and the solenoid plunger would pump in and out. The basic reciprocating motor mechanism was complete.

Recall that the limits of motion of the plunger are prescribed by the solenoid's stroke. In the case of my solenoid, that number turned out to be about ½-inch, meaning, from the point where the plunger was fully inserted to the point where it was fully extended, that travel amounted to half of an inch.

When the crank is thrown to its position where it most closely approaches the solenoid, the crank, through the connecting link, forces the plunger all of the way into the solenoid. Alternately, when the crank is thrown to its position where it is farthest from the solenoid, the plunger is withdrawn to its point of maximum withdrawal. Since we already said that the stroke of the solenoid was limited to ½-inch, and the plunger is linked to the crank, the throw of the crank must also be limited to ½-inch. Thus, the length of the crank arm must be half that, or ¼-inch.

A different make or model of solenoid is likely to have different attributes, including a different stroke. In any case, I would fashion a crank arm whose length is no greater than one-half of the solenoid's stroke. Understand that when I refer to the "length" of the crank arm, I'm not talking about its external dimensions. What I am referring to is the distance between the center of the collar affixed to the crankshaft, and the center of the collar that links the crank to the connecting link. The external dimensions of the crank arm are not relevant.

Engine Timing

As discussed earlier, proper motor operation depends upon precise timing. By this, I mean that the solenoid must be activated, and also deactivated according to the position of the crank and the solenoid plunger. In this motor, the solenoid is turned on and off with a simple leaf switch. The switch itself is actuated with a cam.

***Figure 2-8**: A partially dismantled leaf switch. Metal leaves fitted with switch contacts are clamped in a sandwich comprised of insulating tubes and plates.*

Leaf switches are composed of two or more strips of springy metal fitted with metal switch contacts. At one end the metal strips, or leaves, they are spaced and kept insulated from one another by one or more insulating plates which are inserted between them. Additional insulating plates and tubes, held together with screws and nuts, form a sandwich which completes the switch assembly. A partially dismantled leaf switch can be seen in figure 2-8.

A wire is attached to each leaf. The free ends of the metal leaves comprise the business end of the switch. If you apply a force to the leaves, they will flex. Apply enough of a force to cause them to touch one another, and the switch turns on.

Leaf switches are sometimes composed of many, many leaves, which are actuated in tandem by a toggle or bat handle. In the case of the Peewee Motor, I inherited a damaged switch in which most of the leaves had been burned and destroyed by a short circuit. I dismantled the switch, extracted two good leaves, along with some of the insulating hardware, and built a basic two-leaf switch.

It is a simple matter to fabricate a leaf switch from scratch from bits of springy metal. Hobby brass or shim stock would probably work well. Consider insulating materials like bits of Formica, plastic, acrylic, or fiberglass board, cut from a scrapped circuit board. Another option would be to find a momentary-contact switch with a paddle or lever arm on it. These are often used as interlock switches on the lids of printers or the doors of certain appliances.

My leaf switch was mounted to the L-bracket, near the crankshaft, using a piece of aluminum L-channel and a couple of 6-32 bolts and nuts. Your mounting method may vary, but note that it is advantageous to allow for some adjustability with regard to the switch's exact position.

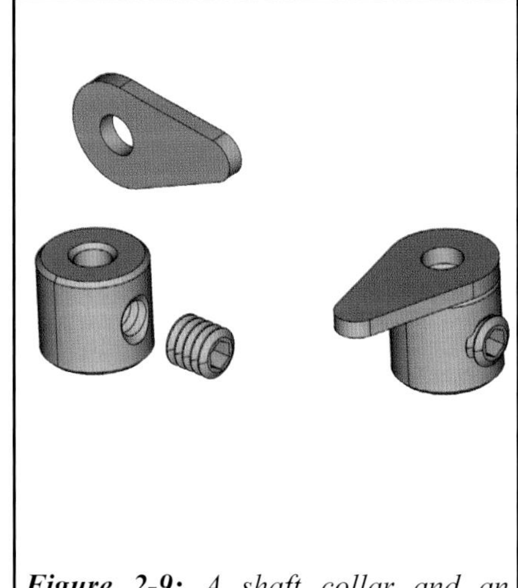

***Figure 2-9:** A shaft collar and an eccentric are soldered together to produce the Peewee Motor's timing cam.*

The switch is actuated with a cam. In the Peewee Motor, the cam was fabricated from a shaft collar, like the ones used to build up the crank, and a short length of brass strip. The strip was soldered to the face of the collar. The strip was dressed with a metal file so that one end matched the circumference of the collar. The other end was allowed to protrude beyond the radius of the collar, though it was also dressed to remove squared corners and to give it a graceful, rounded profile. See figure 2-9.

The cam was installed on the crankshaft between the bearing and the crank, and the switch was adjusted so that, when the cam swept against the leaves, they were depressed, and the circuit through them was completed. The proper fitting of the switch and cam is largely a trial-and-error proposition. The lobe projecting from the cam should not be excessively tall, as this is likely to place needless drag on a motor that is not particularly powerful to begin with. On the other hand, it must be of sufficient height to positively actuate the switch you are using. In my case, a lobe height of about 0.125-inch was sufficient.

The last and most critical step is to set the motor's timing. To do this, I loosened the set screw on the cam just enough to allow me to twist and adjust its angular position on the crankshaft, but left it tight enough so that the cam would retain its position when I let go.

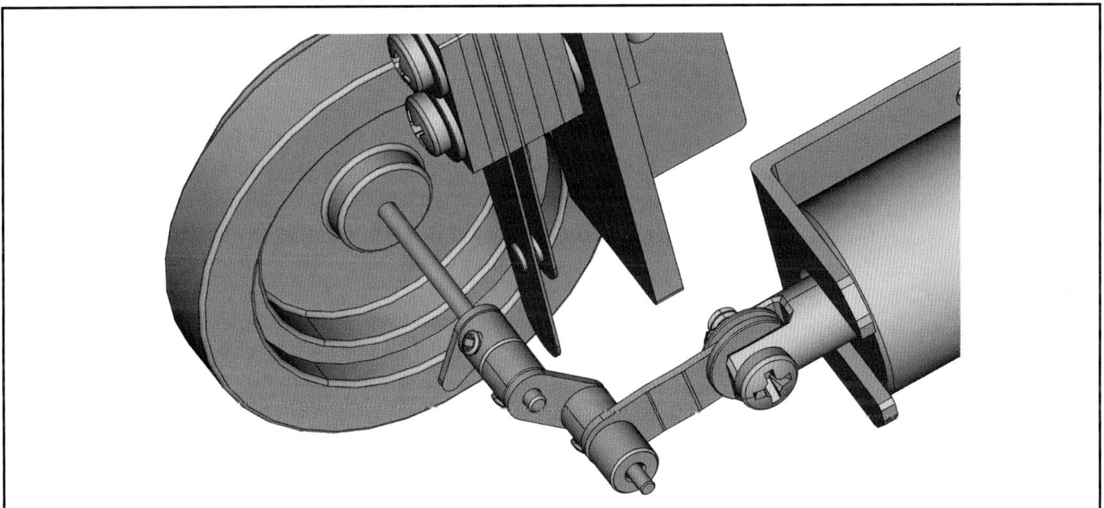

***Figure 2-10**: A view of the crank, cam, and reed switch. For clockwise rotation, the cam leads the crank by roughly 90 degrees. See text for details.*

Motor timing is best expressed in terms of degrees of rotation. The Peewee Motor was set up to run in a clockwise fashion, as viewed by the crank end of the shaft. The position of the crank at which the plunger is fully inserted into the solenoid coil is considered "top dead center", or zero degrees. In the case of the Peewee Motor, this means the crank is pointing to the right.

Angular displacement increases as we rotate the crank clockwise. When it points down, we've reached the 90 degree position, to the left is 180 degrees. When the crank points upward, that's the 270 degree position. When we've completed a revolution of a full 360 degrees, we've returned to the top-dead-center, or 0 degree, position. Figure 2-10 shows the general relationship between the crank, cam, and reed switch.

To time the motor, I rotated the crank until it was positioned vertically, at the 270 degree position. Then, I adjusted the cam so that the leaf switch had just closed. Next I rotated the crank clockwise toward the top dead center position. I verified that the switch opened back up, just before the plunger reached the point at which it was fully inserted into the coil.

Again, adjustments here require some tinkering. The point at which the switch will close is determined by the angular position of the cam on the crankshaft. The amount of time the switch will remain closed, the dwell time, depends upon the profile of the cam, and to some extent, the height of the cam's lobe. In any event, once all of the adjustments and modifications were made, I tightened the set screw on the cam to lock it into place.

Testing and Results

The motor was wired according to figure 2-11. Stated simply, the leaf switch was wired in series with the solenoid, and the leads were terminated with a pair of "banana"-style binding posts, one red, and one black. The binding posts provide a means by which a direct-current power supply can be connected to the motor, and are visible at the top of the motor in figure 2-1.

By convention, the power supply's positive lead is connected to the red binding post, while the negative lead is connected to the black. In truth, inverting the polarity of the power supply (swapping the positive and negative leads) has no effect on this motor whatsoever. However, in other motors described in this book, a miswired supply can cause the motor to run in reverse or not at all, so this is a detail worth paying attention to.

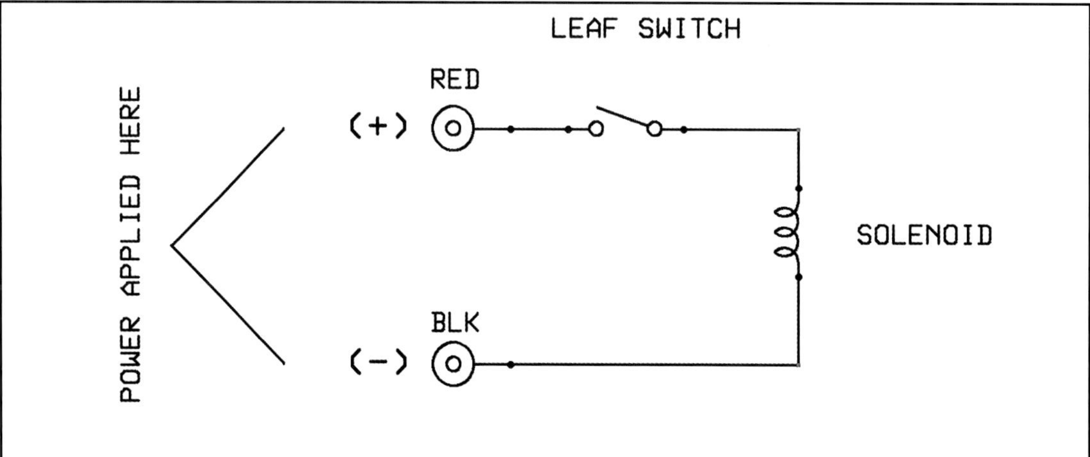

***Figure 2-11**: The wiring for the Peewee Motor is simple. The solenoid and leaf switch are connected in series and then terminated with binding posts.*

Because this motor is of the single "cylinder" reciprocating type, most of the time it is not self-starting. The flick of a finger against the flywheel will set the machine into motion. Once in operation, it is entertaining to watch, as the reciprocating parts make interesting noises when in operation.

The solenoid used to build the Peewee Motor has a label specifying an operating voltage of 24V, D.C. I have run the motor on as little as 8V, D.C., and as high as 30V. Figure 2-12 shows the relationship between motor speed and applied voltage. Notice that the motor tops out around 1000 RPM, after which further increases in voltage do little to increase the motor's speed. This is no doubt the point where the motive force of the solenoid on the crank is offset by parasitic losses in the machinery, including friction and windage.

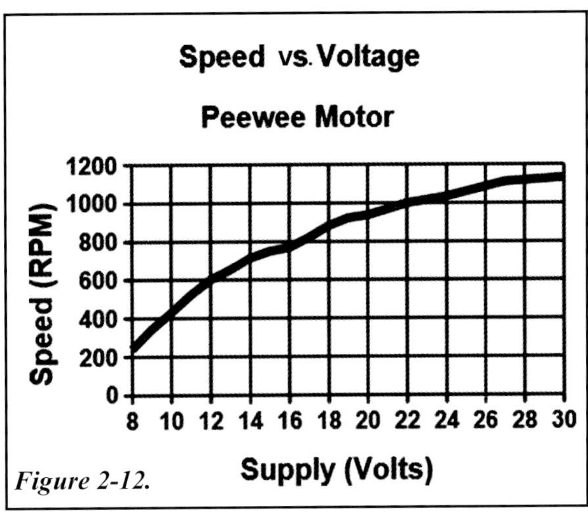

Figure 2-12.

Critical Considerations

Motors of this type will run in only one direction. If you wish to reverse the direction that the motor runs, you need to change the position of the cam. Try loosening the set screw, and rotate the cam 180 degrees from its present position. Then tighten the set screw again. You may need to do some fine tuning.

The crank represents an unbalanced load, because it extends asymmetrically from the shaft. Unbalanced loads on rotating parts lead to vibration. Consider, now, that the crank does not operate in isolation. It's linked to the solenoid plunger, which reciprocates. The force of the magnetic field in the solenoid accelerates the plunger from a dead stop, and draws it into the coil, where the plunger then decelerates and comes to a halt. The inertia of the flywheel and the action of the crank then accelerates the plunger in the opposite direction, drawing it out of the coil, where it is decelerated to return to its starting position. At 1000 RPM, this violent sequence must take place in less than six one-hundredths of a second. The connecting link, being attached, is doing likewise. My point is that the sum total of all of this imbalance and thrashing to and fro results in a lot of vibration.

In practical reciprocating engines, some of this vibration is suppressed through the use of counterweights which are typically attached to the crankshaft or the crank itself. The weights are sized, shaped, and positioned to generate intentional forces that neutralize, to a large extent, the vibration described above. The Peewee Motor would probably benefit from a counterweight. One way to implement this would be to drill and tap some holes in the face of the flywheel, where screws could be used to attach washers.

When coils are energized and magnetic fields are created, energy is actually stored in that magnetic field. If we disconnect power to the coil abruptly, the field collapses, and the stored energy has to go someplace. Some of the energy is dissipated as heat, while some reemerges as a voltage "spike," more properly termed a counter-EMF (Electro-Motive Force). Evidence of this appears between the contacts of the leaf switch in the form of bright, hot sparks. Discharges of this type can be rough on switches and will eventually erode their metal contacts.

This discharge can be accompanied by fairly high voltages. While I would not expect the magnitude of these voltages to be hazardous to anyone in reasonable health, touching the switch or solenoid contacts while the motor is in operation may result in an unpleasant electric shock.

Another aspect of the Peewee Motor that could be improved has to do with the angular relationship between the crank, the connecting link, and the plunger. Developing this thought requires a brief detour in subject matter, but I think you will find the trip worthwhile.

Vectors and Vector Math

Humans like to quantify things, so we use numbers to represent the magnitude of the measurements we make. Thus, we might speak of a two-liter soda bottle, the 32-degree-Fahrenheit freezing point of water, or one bushel of grain. These are called scalar quantities.

Sometimes, however, simply expressing a quantity in terms of a number is not enough. It might be correct to say that the velocity of a bullet fired by a .45 automatic pistol is 1,060 feet per second, but I submit that it is essential to know whether that bullet is flying towards you or away from you! Some quantities, then, are best expressed by two dimensions, a magnitude and a direction. Numbers of this type are called vectors.

The effects of multiple vectors can be added together, just as scalar quantities can be. The math is just a little bit more involved. Consider a simple example of vector addition that involves a river, which flows from west to east at a rate of 3 miles per hour. A man in a rowboat starts at the south shore of the river and rows due north at a rate of 1 mile per hour. The situation is depicted in figure 2-13. How fast, and in what direction, does the man's boat actually travel?

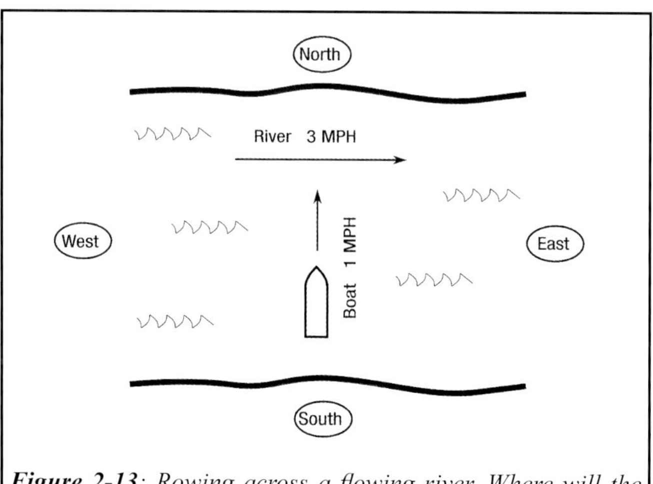

***Figure 2-13**: Rowing across a flowing river. Where will the boat end up?*

The first step to solving this problem is to redraw it as a vector diagram. The vectors are represented as little arrows, which are each drawn to point in the direction specified in the problem statement. The length of each vector in the drawing is proportional to the magnitude of the velocities expressed in the problem. Thus, vector "a," the vector representing the speed and direction the boat is being rowed, is drawn as an arrow one unit long, pointing north (upward.) Vector "b", the vector representing the velocity and direction of the flowing river, is drawn as an arrow three units long, pointing east (to the right.)

Since the velocity of the river and the velocity of the oarsman in the boat are additive, the velocity vectors are also added. Graphically, this means aligning them so that the point of vector "a" touches the tail of vector "b." The result, vector "c", is the vector that joins the tail of "a" to the point of "b." See figure 2-14.

The answer to this problem is probably what you'd have guessed through intuition, namely, that the rowboat drifts to the right as it crosses the river, defining a path that moves in a north-easterly direction. But can we calculate actual values? Certainly, we just need a little bit of trigonometry knowledge and a halfway-decent calculator. Note that in setting up our vectors we created a right triangle. A right triangle is one in which one corner measures 90 degrees.

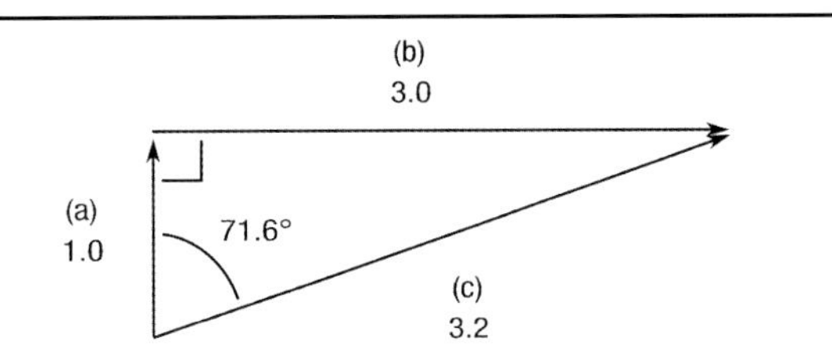

***Figure 2-14**: The rowboat problem resolved with vectors. The boat will travel 3.2 miles per hour and veer 71.6 degrees.*

The Pythagorean theorem (figure 2-15) can be applied here.

$$a^2 + b^2 = c^2 \qquad c = \sqrt{a^2 + b^2}$$

***Figure 2-15**: The Pythagorean theorem.*

So, we square the magnitude (length) of vector "a", square the magnitude (length) of vector "b", add these two values together, and then take the square root of the result. The magnitude of vector "c" is therefore 3.2 miles per hour, meaning that the speed of the boat along the path "c", as the result of the rowing combined with the river movement, is actually 3.2 miles per hour.

The boat's direction, the angle theta, can be calculated using the definition of the tangent function shown in figure 2-16.

$$\tan\theta = \frac{opposite}{adjacent} \qquad \theta = \tan^{-1}\frac{opposite}{adjacent}$$

Figure 2-16: *The trigonometric function called Tangent.*

The side opposite angle theta, the angle we're interested in, is vector "b," which measures 3 miles per hour. The adjacent side of theta is vector "a," measuring 1 mile per hour. If you divide three by one and then take the inverse tangent of that value, you end up with 71.6 degrees. The path of the boat, then, is 71.6 degrees to the east of your intended path.

Since more than one vector can be combined to produce a single vector, it shouldn't come as a surprise that a single vector can be decomposed into its component vectors. The next example shows how this can be done.

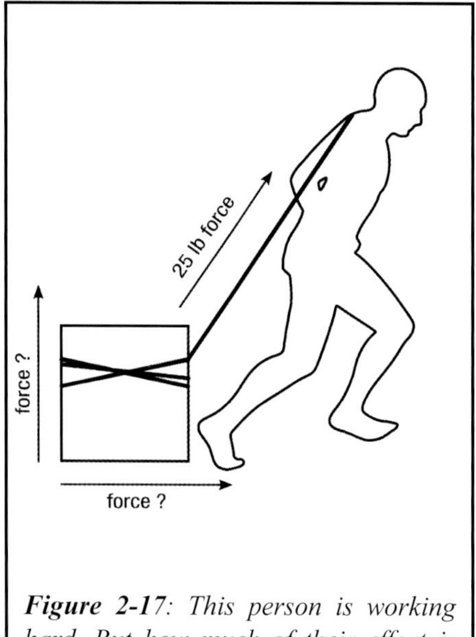

Figure 2-17: This person is working hard. But how much of their effort is actually going towards the objective?

It seems that in a factory somewhere, workers are asked to drag heavy wooden boxes across a concrete floor. Each box has a tow rope, which is pulled by the worker. In this example, the rope is being pulled by a comparatively tall person, such that it rises at an angle of 60 degrees with respect to the floor. Yanking on the rope he exerts a pull of 25 pounds. How much of that force is actually being applied toward the objective of moving the box? See figure 2-17.

Thoughtful consideration of the problem suggests that the rope actually applies two forces in perpendicular directions. Yes, force on the rope will translate into some amount of horizontal force which will act to slide the box. On the other hand, some amount of the rope's force will be translated into a vertical lifting force.

Analysis begins, once again, with a vector diagram. See figure 2-18. Vector "c" represents the direction and magnitude of the force on the rope. Vector "a" represents the force and direction of the desired horizontal sliding force, and vector "b" represents the unwanted vertical lifting force. Angle theta is known, as it is specified as 60 degrees.

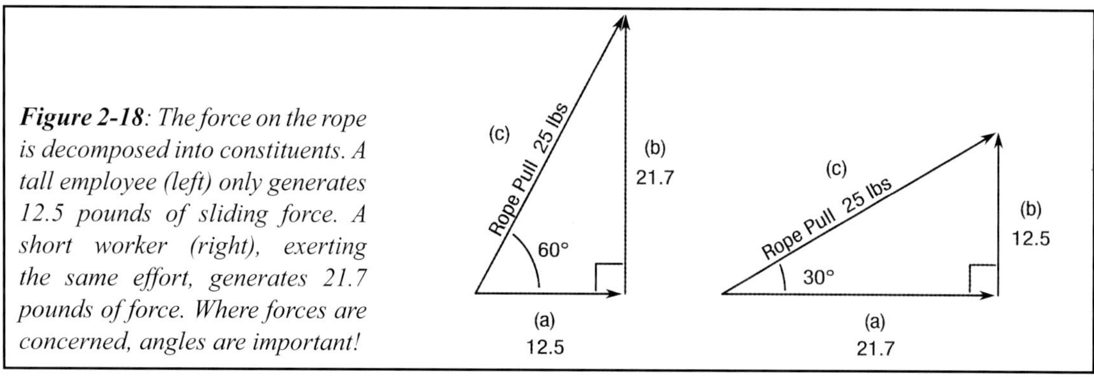

***Figure 2-18**: The force on the rope is decomposed into constituents. A tall employee (left) only generates 12.5 pounds of sliding force. A short worker (right), exerting the same effort, generates 21.7 pounds of force. Where forces are concerned, angles are important!*

Vector "a" is considered the adjacent side of the right-triangle formed by vectors "a," "b," and "c". To solve for the magnitude of vector "a," we can use the Cosine identity in figure 2-19.

The cosine of 60 degrees times the hypotenuse (which is vector "c") equals 12.5 pounds. Interesting... the worker is applying 25 pounds of force to the rope, but only 12.5 pounds of force is acting to help slide the box! Where is the rest of his effort going? Vector "b," the opposite side of the right triangle, represents the vertical or lifting component, so lets use the sine identity in figure 2-20 to solve for "b".

The sine of 60 degrees times the hypotenuse (which is vector "c") equals 21.7 pounds in the upward direction! Wow! Most of the force being applied to that rope is going to no useful purpose!

$$\cos\theta = \frac{adjacent}{hypotenuse} \qquad adjacent = (hypotenuse)(\cos\theta)$$

***Figure 2-19**: The trigonometric function Cosine.*

Now let's give the job to a different worker, one who is substantially shorter than the first guy. The new worker is so short that, when he holds the rope, it only makes a 30-degree angle with respect to the ground. Let's assume that he is physically capable of applying the same 25 pound force on the rope. If we run through the calculations again, we find that the short fellow directs 21.7 pounds toward sliding the box, and only 12.5 pounds in the vertical direction. All we changed is the angle at which the rope was towed, and yet the magnitude of the forces applied horizontally and vertically reversed themselves!

By decomposing the vector representing the pull on the box's rope, we can demonstrate mathematically that the force applied by the tall worker translated poorly into the desired horizontal force, while the force applied by the short worker did much better.

$$\sin\theta = \frac{opposite}{hypotenuse} \qquad opposite = (hypotenuse)(\sin\theta)$$

***Figure 2-20**: The trigonometric function Sine.*

Torque

The word torque is commonly used to describe the twisting force delivered to an object. A torque value may be used to describe how "tight" a bolt or nut should be fastened, or how aggressively the shaft of a motor, like the Peewee, can drive a rotational load.

Torque, represented by the Greek letter tau (**τ**), is produced when force is applied to the handle of a wrench, lever, or crank. It's defined as the product of that force times the length of the "lever arm" through which the force is applied. Torque is therefore measured in units of force-times-length, so common units you're inclined to see might include ounce-inches, pound-feet, or in the metric system, newton-meters. See figure 2-21.

$$\tau = Fd$$

***Figure 2-21**: The formula for torque is the product of force times distance.*

Figure 2-22 depicts a large wrench affixed to the head of a bolt. The wrench is 2-feet long. A force of 10 pounds is applied to the wrench's handle. The result is 2 by10 or 20 pound-feet of torque. Note that if the same amount of handle force were applied to a 4-foot wrench, this would double the torque produced. This mathematical relationship is why, in the face of a recalcitrant lug nut, some mechanics will employ the tactic of extending the lug wrench's handle with a piece of steel pipe.

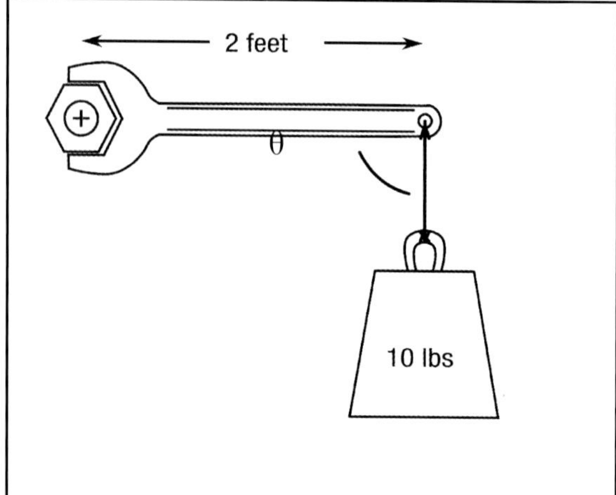

***Figure 2-22**: Torque on a bolt head is the product of the force applied times the length of the lever arm. Angle theta is presumed to be 90 degrees, but if not, this has to be accounted for.*

By definition, the force is always applied perpendicularly (meaning, 90 degrees with respect) to the lever arm. If we apply the force at some other angle, the resulting torque is diminished. This can be demonstrated by decomposing the applied force vector into two constituent vectors, one in the direction of the crank handle, and one perpendicular to it. If we remember that changing the angle of a tow rope attached to a box will change the amount of pulling force on the box, it makes sense that changing the angle of the pulling force on a wrench handle might affect the amount of torque produced.

Because of this, the full expression for torque includes a sine theta factor (figure 2-23):

$$\tau = Fd \sin \theta$$

***Figure 2-23**: A torque formula that accounts for the angle of the applied force.*

Angle theta represents the angle of the applied force with respect to the lever arm. If we apply the force in a perpendicular fashion, as is ideal, theta is by definition 90 degrees. The sine of 90 degrees is one, which is then multiplied by the force times the length of the crank arm to give you torque as defined earlier. On the other hand, if the force is not applied perpendicular to the crank, theta becomes some number other than 90 degrees. The sine of a theta value greater or less than 90 is a fractional value, some number less than one. When multiplied by the force and length numbers, the torque value becomes less than what it would otherwise have been.

Cranks, Links, and Angles

If you spin the flywheel of the Peewee Motor with your fingers and watch the behavior of the crank and connecting link, it immediately becomes apparent that the angle between the crank arm and the link is constantly changing. This means that the capability of the solenoid to impart torque on the crankshaft is also constantly changing. The way these angles change with respect to one another has to do with the length of the crank arm and the length of the connecting link.

Out of curiosity, I wrote a set of equations to study torque as a function of the connecting link length. Initially, my equations assumed a crank arm of one unit in length, a connecting link of one unit in length, and an hypothetical solenoid fired over the 90-degree period between the 270-degree mark and top dead center. I then plugged in several different values for the length of the connecting link and plotted the data to see how torque is affected as the connecting link's length changes. The results are summarized in figure 2-24.

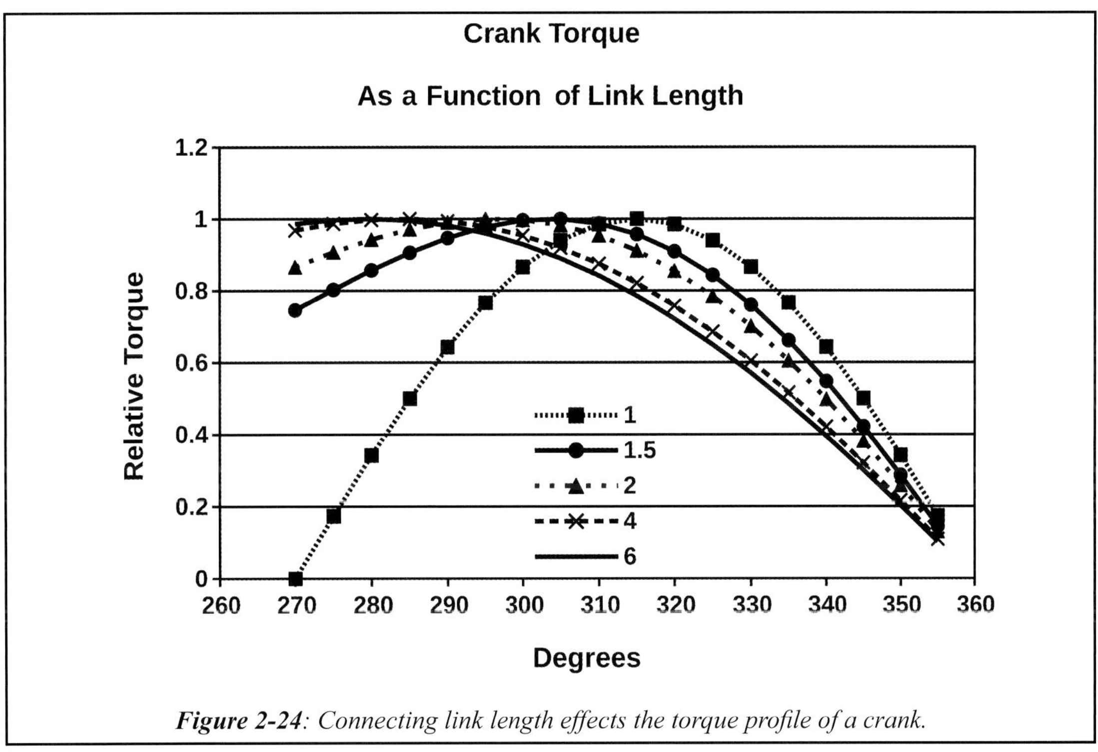

Figure 2-24: Connecting link length effects the torque profile of a crank.

If we assume that the force produced by the solenoid is uniform over its entire stroke, the most efficient conversion of force-to-torque seems to occur when the connecting link is in the neighborhood of 1.5 to 2 times the length of the crank arm.

All that said, it bears mentioning that the force produced by real solenoids is typically very non-uniform. Figure 2-25 (adapted from Underhill) represents data collected on an experimental solenoid, similar in features to the solenoid used in the the Peewee motor. This graph depicts the relative force on the plunger as a function of it's stroke.

Predictably, 100% force is available only at 100% insertion, and the force on the plunger falls significantly as the plunger is withdrawn. However, the mean available force (dotted line) is above 60% over the entire stroke and even at full withdrawal, a respectable 30% of the attractive force is still available to do useful work.

If one were truly committed to optimizing the design of a Peewee-like motor, one might considering making a series of measurements to characterize the actual solenoid used. This data could be plotted against a set of curves like those depicted in figure 2-24, to determine which connecting link length would work to the best advantage given the irregularities in the plunger force over its stroke.

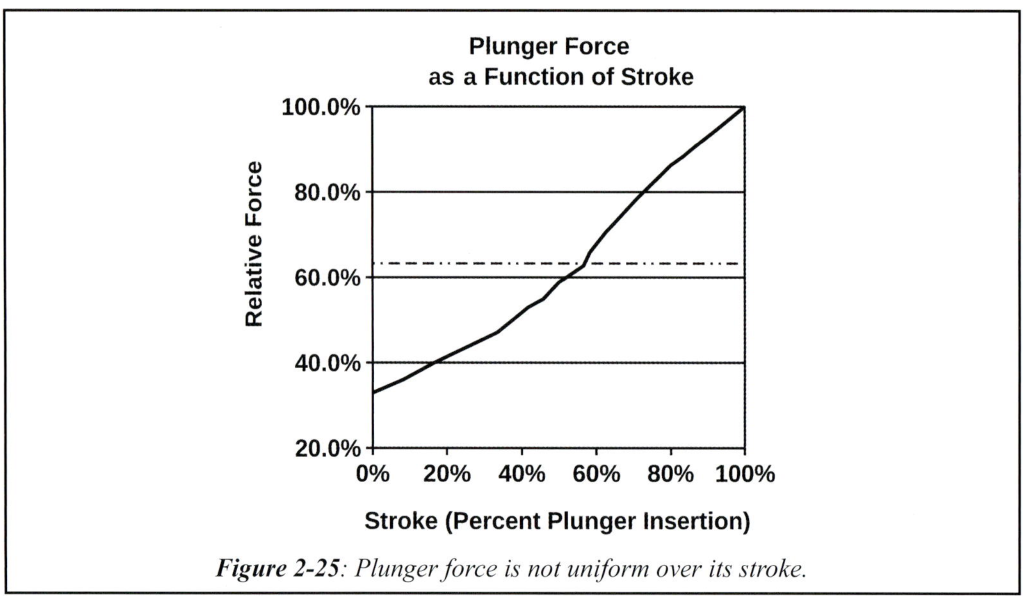

***Figure 2-25**: Plunger force is not uniform over its stroke.*

Chapter III
The Texas Motor

***Figure 3-1**: The Texas Motor, front.*

Amateur Radio is a wonderful and instructive pastime. There is great fun in exercising the privilege to build, experiment with, and use radio transmitting equipment. Over the years, I've learned an awful lot, and met a lot of nice people. The ham radio fraternity is one immersed in a rich culture with longstanding traditions. One such tradition is the so-called "hamfest".

Hamfests are social gatherings where hams converge to buy, sell, and trade parts, equipment, and tall tales. Hamfests may be conducted in a community center, the conference room of a hotel, or in an unoccupied school building on a Saturday, but where I live, the weather is nice most of the time. Most of the hamfests I've attended have been conducted outdoors, in a parking lot or in an open field.

One never knows what treasure may be found at a hamfest. Sellers cover their truck beds, tailgates, and folding tables with vintage radio transceivers, old tubes, antenna parts, recording equipment, computer parts, switches, wire, decommissioned military electronics, and who knows what else. Prices are often good, and many sellers like to haggle. It's great fun for anyone who tinkers with electricity or electronics, ham operator or not.

On one occasion, I had visited a hamfest out in Sierra Vista, Arizona. I was returning to my parked car when I passed a pickup truck with three men in the back. They were hunched over some old equipment, and were busy pulling vacuum tubes out of it.

"*Looks like you found something good,*" I said offhandedly, honoring the ham tradition of friendliness. "*We got all of this stuff cheap,*" one man answered. "*We're only interested in the tubes, though*". A light lit up in my head. "*What are you gonna do with what's left?*" "*Dumpster, I guess,*" the man answered. "*Could I have it?*" I asked. "*It's all yours!*"

One of the chassis I salvaged from these guys was part of an old reel-to-reel tape recorder. When I got home, I took it apart and extracted fistfuls of useful parts likes nuts, screws, springs, pulleys, and similar mechanical odds and ends, not to mention a box full of electronic components. I also extracted a beautiful flywheel, perhaps 3-½-inches in diameter, along with its matching bearings. That flywheel begged me to become part of a Marvelous Magnetic Machine, and eventually it did.

The end result is machine I call the Texas Motor. See figures 3-1 and 3-2. Like the Peewee, this is also an attraction motor, which relies on a crank to convert reciprocating to rotary motion. The Texas Motor differs from the Peewee in that motive force comes from the action of fixed-core coils upon a steel armature. Its most distinguishing feature is its "walking beam," a teeter-totter-like mechanism that couples the armature to the crankshaft. The Texas Motor bears more than a passing resemblance to the pumpjacks one might see in the oilfields of Texas, and so this is where the motor gets its name.

Figure 3-2: *The Texas Motor, rear.*

The Base

In searching for parts from which to build the motors in this book, I identified all of the second-hand and thrift stores near my house and made a habit of visiting them with regularity. Among the things that one tends to find in these stores is an endless variety of brass decorative items. There are brass vases, candle holders, ashtrays, incense burners, urns, bowls, boxes, and statues. During one visit, I found a beautiful solid-brass trivet. I purchased it for two dollars, and it ultimately became the base for the Texas Motor.

The base is square, roughly 6-inches on a side, and about 1-inch tall. It's heavy, which is a benefit in regard to stability. The brass is thick, a benefit in regard to strength, and it has an attractive form and polish, which gives the finished motor an air of precision and craftsmanship. Brass is a wonderful material to work with, because, despite its relative strength, it's extremely easy to cut, drill, file, or tap with simple hand tools.

Bearings and the Flywheel

As alluded to in this chapter's introduction, the flywheel and bearing set used in the Texas motor was extracted from a reel-to-reel tape deck. In form and function, they differ little from the cassette flywheel and capstan described in the last chapter, with the exception of their much larger size.

Since the Texas Motor is a horizontal-shaft motor, like the Peewee Motor, it follows that a vertical surface was needed in which to mount the top-hat-shaped bearing housing. In the Peewee, this vertical mounting surface was provided through the use of an L-shaped bracket. The bracket was sufficiently tall that it could be mounted directly to the base, and when the bearing was mounted to the vertical portion of the bracket, the flywheel cleared the base and could rotate without interference.

Figure 3-3*: The L-bracket sits on aluminum spacers and is bolted to the base. The bearing fits into the L-bracket as indicated. Fasteners are not shown for the sake of clarity.*

In this motor, the L-bracket was composed of a 3-½-inch length of 2-inch aluminum angle. To assure that the flywheel would clear the motor's base when the bearing was installed, the L-bracket was fitted with four legs, giving it the appearance of a kitchen chair. The legs are aluminum spacers, 1.875-inches in length. The L-bracket and its legs were anchored with 8-32 nuts and threaded rod, which pass through four holes drilled in the base. See figure 3-3.

I drilled a large hole in the vertical face of the angle, big enough to allow for the insertion of the top hat-shaped bearing. I drilled three more holes to match the holes in the bearing's mounting flange, and secured the bearing to the L-bracket with three 6-32 screws and nuts.

At some point, I concluded that the capstan shaft on the flywheel was too short for my purposes, and that I'd have to replace it with something. The flywheel was secured to the capstan shaft with a set screw. I loosened the screw, and the capstan was easily removed. I then measured its diameter, hoping I could find a similar but longer shaft elsewhere.

The reel-to-reel recorder from which these parts were salvaged was a Japanese-made model. The fasteners and internal components had metric measurements. This included the capstan. Its diameter turned out to be 8 millimeters, a source of concern to me because metric rod is not the sort of thing one typically finds in a hardware store in the U.S.A. I began to ponder where I might find the shaft I needed.

My problem was quickly solved when a friend gave me an old dot-matrix printer that he had rescued from the trash. Like tape recorders, old printers are another excellent source of interesting electromechanical parts like stepper motors, gears, and metal shafts. Because printers are virtually all manufactured outside of the United States, their internal parts tend toward standard metric sizes. The matrix printer I just mentioned yielded a nice long section of polished 8-millimeter shaft—just what the doctor ordered.

I fit the new shaft into the flywheel and tightened its set screw. This locked the flywheel onto the shaft. Then, I inserted the shaft into the bearing, already mounted in the L-bracket. I gave the flywheel a twist and it spun like a top. A drop of sewing machine oil on the shaft lubricated the bronze inserts in the bearing, and the flywheel turned even more smoothly.

The Crank

The Texas Motor is another reciprocating engine, so it needs a crank to convert the linear force, generated by the electromagnets, into rotary motion. Now, there's no reason why I couldn't have used the same basic crank design I developed for the Peewee motor. On the other hand, the point of this book, at least in part, is to demonstrate the engineering opportunity that can be found in seemingly unrelated odds and ends. The crank used in the Texas Motor emerged as an adaptation of an old radio instrument knob. Its features bear description, because they were exploited to the advantage of the resulting motor.

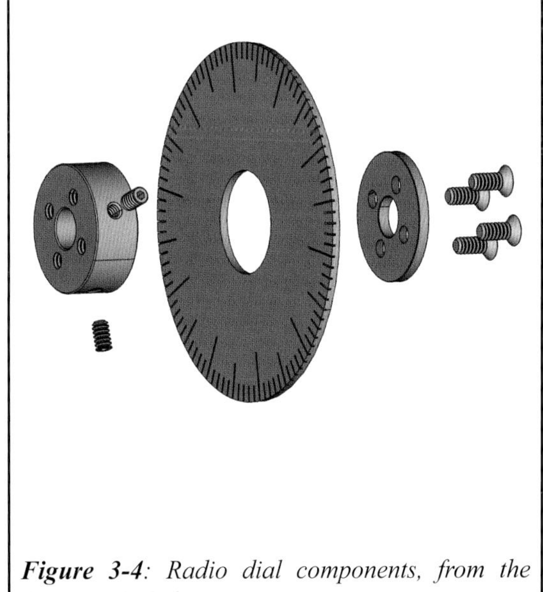

***Figure 3-4**: Radio dial components, from the front, exploded.*

The knob was a large Bakelite plastic affair. Attached to the knob, with four screws and a small retaining plate was a dial, an aluminum disk etched with graduations and numbers. I removed the screws, and the dial came off. I noticed, then, that the knob was not solid plastic.

Rather, it was composed of a cylindrical aluminum core, which had been molded into the interior of the knob. I carefully squeezed the knob in a vise to fracture the plastic, and extracted the core. The core had two set screws to lock it to any shaft it might have been installed on. In this sense, it was not unlike one of the shaft collars I introduced in Chapter 2, only much, much larger — about 1-inch in diameter. Figure 3-4 depicts the parts that were extracted from the instrument knob, looking from the front.

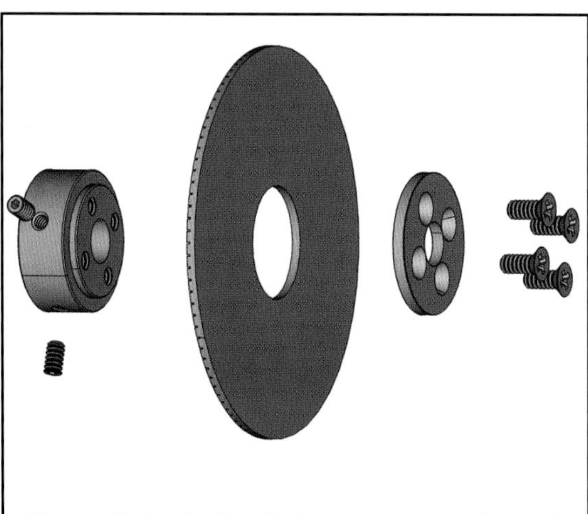

***Figure 3-5**: Radio dial components, from the rear. Note how the graduated dial plate fits to the shoulder of the core and is clamped in place by the retainer plate and screws.*

The rear face of the core, the end to which the graduated dial had been affixed, had a stepped contour or "shoulder" cut into it. This diameter, somewhat less than the 1-inch outside diameter of the cylinder, fit into a large hole in the center of the aluminum dial plate. The dial screws, when tightened, forced a retaining plate against the graduated dial plate, effectively clamping it to the shoulder on the cylinder. This relationship is more evident from Figure 3-5.

Two modifications rendered the core suitable for use as a crank. First, I drilled out the bore of the core to 8 millimeters, so that it would slide onto the crankshaft of my motor. Next, I noted that the holes into which the dial-retaining screws had been threaded had been bored and tapped all the way through the cylinder from one face to the other. To create a crank handle, I located a 6-32 bolt, a couple of washers, and ¼-inch diameter steel standoff. I inserted the bolt through the washer and the standoff, and then threaded the end of the bolt into one of the holes in the aluminum cylinder. This resulted in a crank with a ½-inch throw.

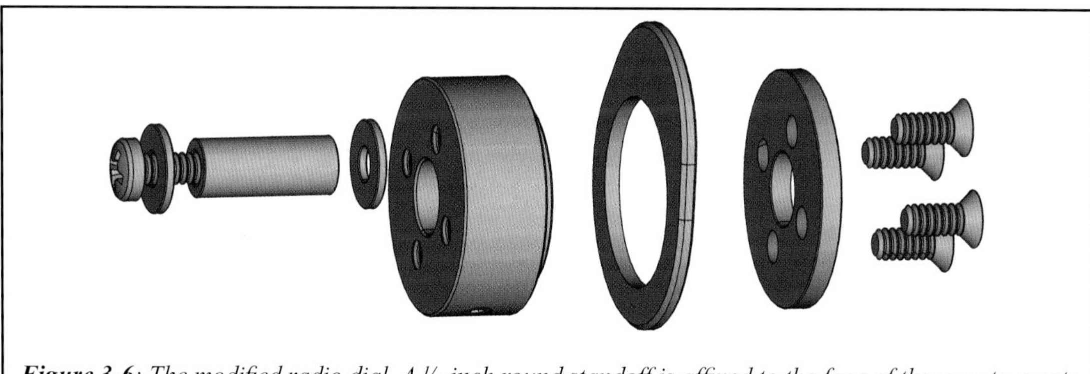

***Figure 3-6**: The modified radio dial. A ¼-inch round standoff is affixed to the face of the core to create a crank. The graduated dial plate has been cut and filed to an eccentric shape to form a cam.*

Further, I reasoned that if I cut down the graduated dial plate, filed it into an eccentric, and then reinstalled it on the aluminum core as the dial had originally been, I could use this as a cam by which to time the motor's firing. The fact that the cam was clamped into place would mean that the angle between the cam and the crank could be changed simply by loosening the screws in the retaining plate, and that the angle between the cam and crank was infinitely and precisely adjustable. The modified radio knob hardware, better illustrated than explained, can be understood by studying figure 3-6.

Two additional advantages arise from this design. First, by combining the crank and cam into one unit, the combined assembly is smaller and occupies less space on the crankshaft. Second, it becomes possible to dismantle the motor and remove the crankshaft without upsetting the timing relationship between the crank and the cam.

Admittedly, the odds of another builder finding the same kind of junked radio knob that I used in this adaptation is pretty slim. Nonetheless, there is merit in presenting the design because there are endless alternatives. Other radio or instrumentation knobs may have similar internal parts, or unrelated scrap could be used to replicate the pieces I just described.

Coils and Backstrap

As I've indicated, I like variety in my motors. So where the Peewee Motor makes use of a commercially manufactured, hollow-core solenoid assembly (acting upon a movable plunger), the Texas Motor generates its operating forces with a pair of homemade fixed-core electromagnets which act upon an armature in the form of an attraction plate. The basic form of the coil assembly is U-shaped, and can be seen in figure 3-7.

The base of the electromagnet assembly — the back strap — was cut from a piece of 0.125-inch (⅛-inch) steel strap, procured at a local hardware store. The steel measures about 5-inches in length and 0.75-inch in width. I drilled four holes in it.

Two of the holes were drilled large enough to pass a #8 bolt, and were later used to fasten the back strap to the motor base. The holes are located about 0.375-inch (⅜-inch) from each end of the strap. When bolted to the base, the back strap rests upon a couple of standoffs, which raises it above the surface of the base about 0.625-inch.

The other two holes in the back strap are just over ½-inch in diameter, large enough to pass the ½-inch steel bolts which are used as cores for the electromagnets. Measured from center to center, the holes for these bolts are spaced 1.875-inches apart, and are centered on the strap.

The bolts used for the coils' cores are each 2-½-inches long. Each core is anchored to the back strap with a pair of ½-inch nuts. The heads of the bolts face upward, and become the pole faces for the finished electromagnets.

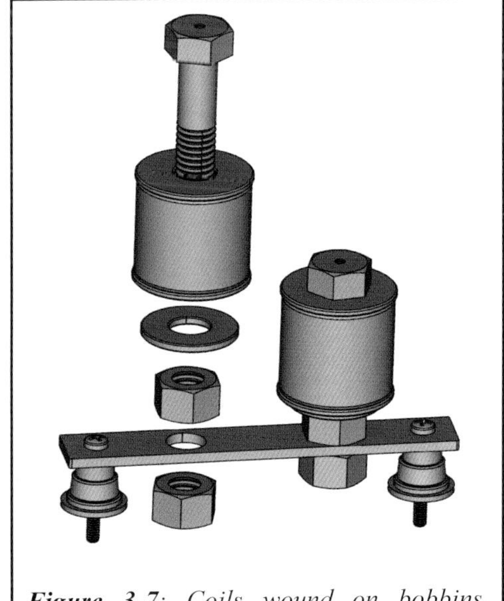

***Figure** 3-7: Coils wound on bobbins, ½-inch bolt cores, and the steel backstrap.*

The wire that comprises the coils was not wound directly on the core bolts. Rather, it was wound on a pair of bobbins. The coils were fabricated first, and then installed on the core bolts just before the bolts were installed on the back strap.

The bobbins were custom-made from paper and stiff cardboard, which I salvaged from some file index cards. A summary of how the bobbins were made is as follows:

I started with a ½-inch bolt, which I used as a mandrel. I wrapped a strip of waxed paper around the threads of the bolt a few times, and secured it with cellophane tape. The purpose here was to increase the diameter of the bolt slightly, and to assure that that bobbin I was about to fabricate would not become stuck on the mandrel.

Next, I cut a strip of printer paper, 1.375-inches wide, of some indeterminate length. Using the wax-papered bolt as a winding tool, and a little bit of glue to hold things together, I wound the strip around the bolt, like a rolled-up-carpet, until I had fashioned a stout paper tube. Note that the width of the paper strip determines the ultimate width of the bobbin.

Once the glue on the tube was dry, the next step was to fashion the flanges or ends of the bobbin. I used a compass to lay out two circles onto a piece of stiff cardboard, and I used scissors to cut them out. These circles measured 1-½-inches in diameter.

Obviously, it is necessary to make a large hole in the center of each flange so that they can be slid onto the paper tube fabricated earlier. I used a compass to draw a circle at the center of each flange, which defined the size of the hole I intended to make. However, instead of simply cutting out the material inside the circle with scissors, I used a hobby knife to make a series of slashes across its diameter. Stated another way, I cut the interior of the small circle into little pie slices.

The reason for cutting the center hole this way becomes more apparent when the bobbin's ends are forced onto the tube I made earlier. The pie slices bend to one side and become little "fingers" which improve the fit of the flange onto the center tube. Glue, applied to these fingers before the flanges are forced into place, forms an indestructible bond, owing to the large surface area between the fingers and the tube.

The material from which the flanges of the bobbin are made must be sufficiently stiff to contain the wire that will be wound on the bobbin later. If the cardboard is too thin or too light, the bobbin will not have the necessary mechanical integrity. I had some concerns about the cardboard I had used, so I elected to double-up the end pieces, so as to make them thicker. I used my compass and scissors to make a couple more disks, but this time, I cut their centers out cleanly, without resorting to clever "pie" cutting. These disks were glued to the faces of the flanges I had already installed. The various cardboard parts that constitute a complete bobbin can be seen in figure 3-8.

Using the process I just described, I fabricated two bobbins for the Texas Motor. In the interests of aesthetics, I gave each one a shot of black spray paint. Then, to render them immune to changes in humidity, and to further stiffen and stabilize the cardboard,

I dipped each bobbin into a can of polyurethane varnish. I let them soak for five or ten minutes, and then withdrew them. I suspended them on short lengths of wire as they dripped, dried, and finally cured. The finished bobbins became very tough, almost plastic-like, and have an appearance very similar to bobbins found on vintage electrical gear.

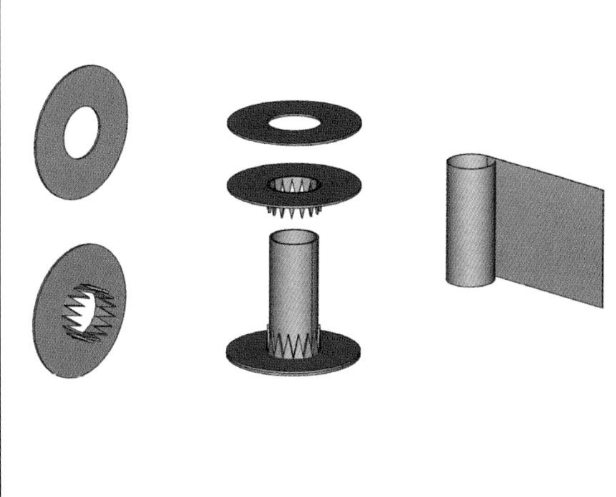

Figure 3-8: *A rolled paper tube and cardboard flanges form the basis for a robust bobbin on which to wind the electromagnets.*

Winding the coils was done in a semi-automated fashion. I took a ½-inch bolt and inserted it through a large fender washer. I slid the bobbin on the bolt, followed by another fender washer, and then a nut. I tightened the nut, and then inserted the free end of the bolt into the chuck of a variable-speed drill motor. It helped to wrap the drill motor in a rag and clamp it gently in a bench vise, because securing the motor in this fashion leaves one hand free to operate the trigger, and the other hand free to guide wire onto the bobbin as it is rotated.

The coil was wound with so-called "magnet wire." Magnet wire is copper wire coated with a thin layer of enamel, which acts as the wire's insulation. Magnet wire is the preferred type of wire for winding electromagnets, because the enamel insulation allows more wire to fit into a given volume than would be possible with wire that is insulated with plastic.

Magnet wire is readily available, both locally and on-line. It can also be found in a wide variety electrical and electronic scrap. Wire can often be harvested from old motors, transformers, television yoke coils, electric clippers, blenders, wall warts, and all kinds of junked appliances. I can't recall where I found the wire that I used for the coils in the Texas Motor, but I can tell you that it was 30 AWG (American Wire Gauge) copper. When I extracted it from whatever it came from, I wound it upon an empty solder spool for storage. When I wound the coils for the Texas Motor, I transferred the wire from the solder spool to the bobbins.

Using a small piece of adhesive tape, I tacked the wire to the bobbin, being careful to leave an 8-inch dangling lead. This lead was wound into a dime-sized roll and taped somewhere out of the way. I then pushed the trigger on the drill motor, and began to wind wire onto the bobbin. I used my hand to guide the wire back and forth, so as to lay down as even a layer as possible.

How much wire did I put on the bobbin? I had no mechanism for counting turns at the time, and in any event, the application is not so critical that a precise count even matters. What I did do was to fill the bobbin with wire to a diameter about 0.125-inch less than the flanges. I estimate this to amount to be around 650 feet of wire for each bobbin. When I got to the end, I secured the free end of the wire with another piece of tape. Incidentally, the resulting coils have a measured DC resistance of 66 ohms (Ω) each.

An electromagnet in this state is quite fragile. Should anything bump against the exposed windings, they can be scratched or damaged. If the leads are yanked on, they can break. If you break the lead at the start of the winding—the one that's buried underneath all the other windings—the only way to reestablish a connection is to unwind and then rewind the entire coil! Needless to say, that's no fun. Thus, when my coils were complete, I wrapped them with a turn or two of fabric medical tape, just to protect the coil windings from scratches or impact.

To strengthen the leads themselves, one has at least two options. Some coil builders cut short lengths of stranded, insulated, hookup wire (something more robust than the magnet wire) and solder these to the magnet wire leads. The solder connections are insulated, and then the body of the coil is wrapped with numerous additional turns of tape. As the tape is being laid down, the fragile magnet wire leads and the insulated solder connections are buried. The hookup wire leads are sometimes woven, over and under successive tape layers, so as to provide strain relief. The idea is to secure the hookup wire leads with the tape so that if they should undergo tension, that force is transferred to the tape, and not to the fragile magnet wire.

In my case, I implemented a variation on the idea. I carefully stripped a long section of plastic insulation off of some hookup wire, which left me with what amounted to a tiny plastic tube. Electronics folks call this "spaghetti tubing". I created two such tubes, one for each of the magnet wire leads on my coils. I threaded the magnet wire through the tube, and then secured the tube with multiple layers of cloth medical tape, as I just described. This is not quite as hardy or abuse-resistant as the previously described technique, but it worked very well for the Texas Motor.

Once the coils where finished, they were fitted onto the ½-inch core bolts and secured with a single fender washer and a nut. A second nut secured each coil/core assembly to the back strap as described above.

The Attraction Plate

In the Texas Motor, motive forces are generated by the interaction between the electromagnets just described, and a part I call the "attraction plate". This might alternately be referred to as the armature.

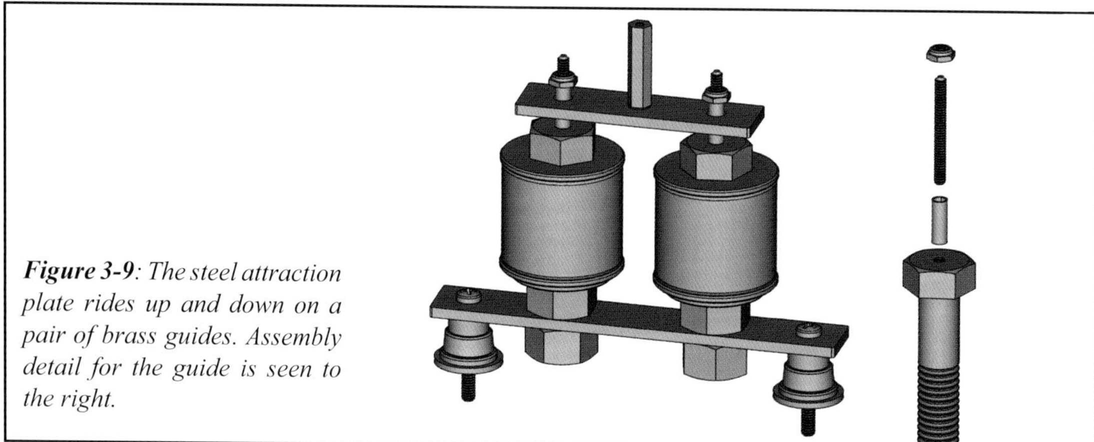

***Figure 3-9**: The steel attraction plate rides up and down on a pair of brass guides. Assembly detail for the guide is seen to the right.*

The attraction plate consists of a 3-inch length of 0.125-inch (⅛-inch) steel strap, 0.75-inch (¾-inch) wide. At the center of the attraction plate is a mounting hole to which I attached a hex standoff, using a 6-32 screw and lock washer. The standoff provides a "handle" or attachment point for a connecting rod, which will be described later.

In theory, nothing more is required in terms of features or attributes. Unfortunately, practical reality introduces problems. The attraction plate functions by being drawn toward the pole faces of the electromagnets, in a path parallel to the axis of the cores—in other words, up-and-down. Yet, there is nothing to restrict the plate from moving laterally. Side-to-side motion gets us nothing. In fact, if the plate moves too far to one side or another, it will begin to move outside of the field produced by the electromagnets, degrading the efficiency of the motor. Clearly, some kind of guide or restraint must be implemented to restrict the plate to vertical motion only.

To answer this, I drilled a hole in the center of each pole face of the motor's electromagnets. Using a 6-32 tap, I cut threads in the holes, and then fitted each with a length of 6-32 threaded rod. These would act as the guides for the attraction plate.

I drilled corresponding holes in the attraction plate, spaced and sized so that the guides could be inserted in the holes. The plate was still free to move vertically, but was now restricted by the threaded guides from moving side-to-side.

The exposed threads of the guides turned out to be problematic. The threads scraped inside the holes in the attraction plate, like tiny rat-tail files, resulting in drag on the motor and needless wear and tear. The problem was easily corrected by covering the exposed threads with 0.625-inch lengths of brass tube. The tubes render the guides smooth, and with them installed, the attraction plate glides up and down effortlessly.

All of this is far simpler to depict than to describe. See figure 3-9. The tubes were secured, by the way, with a couple of 6-32 nuts.

Walking Beam

The walking beam, the teeter-totter-like mechanism at the top of the Texas Motor, is probably the most distinctive feature of this machine. One end of the beam is coupled to the attraction plate with a connecting rod (described later). At the other end, a second connecting rod links the beam to the crank. When the electromagnets are switched on, the attraction plate is drawn downward. The associated connecting rod pulls down on one end of the beam. The opposite end of the beam rises, pulling the crank upward through its connecting rod. Later, the coils are turned off, and the flywheel advances the crank downward. The beam converts this to upward motion at the coil end, and the attraction plate is lifted off the pole pieces.

A curious observer might wonder why the fulcrum or pivot point of the beam is not centered. This asymmetry is purposeful, and to understand the reasoning, we must consider the characteristics of a solenoid-type electromagnet working on a plunger, versus an electromagnet working on an attraction plate.

In the case of the solenoid and plunger, represented by Figure 2-25 in the last chapter, the force declines as the plunger is withdrawn, but maintains a usable amount of force throughout its stroke. The extended working range of the solenoid is the result of the features I mentioned in the Peewee Motor chapter including cone-tipped plungers, coil bores with ferrous end-plugs, and a ferrous frame around the outside of the coil.

An electromagnet and attraction plate is a very different animal. The graph in figure 3-10, also adapted from Underhill, shows the relative force exerted on an attraction plate by a set of experimental electromagnets, and how that force changes as the gap between the plate and electromagnets is increased.

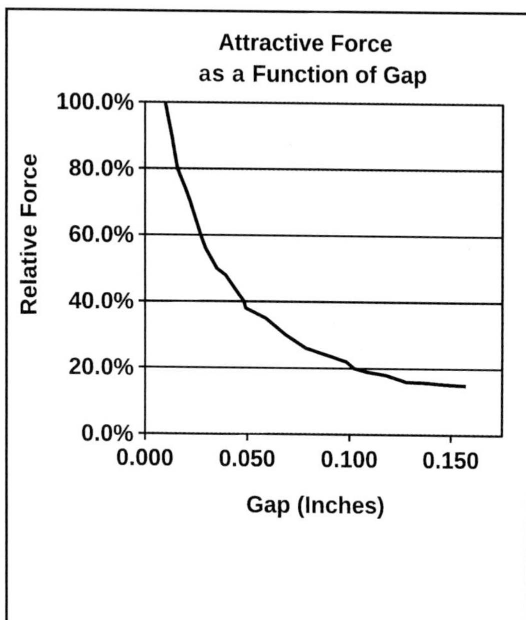

***Figure 3-10**: The attractive force between a coil and metal plate drops rapidly with any increase in gap between them.*

Maximum force is observed when the attraction plate is in actual contact with the coil's core, but as the plate is withdrawn, even a very tiny bit, the force exerted upon the attraction plate drops off precipitously. At a mere ⅒-inch, the attractive force in this particular example has dropped to 20% of its initial value! The takeaway here is that the electromagnet and attraction plate have a very limited working range — in this case measured in fractions of an inch — which translates to a very short stroke.

Now, if the fulcrum of the beam was placed at dead center, the end linked to the attraction plate would swing as far up and down as the end of the beam connected to the crank. Recall that the Texas Motor's crank has a ½-inch throw, which means that the vertical travel of the attraction plate would also be ½-inch. As we've just discussed, this is a comparatively large span over which to influence an attraction plate with an electromagnet.

The idea, then, is to position the walking beam's pivot in such a way that the vertical travel at the crank end is ½-inch, but the travel at the attraction plate end will be much less. By limiting the motion of the attraction plate to a region comparatively close to the coil cores, we maximize their influence on the plate, and thus the efficiency of the motor.

In the Texas Motor, the actual distance between the fulcrum of the walking beam and the point on the beam where the connecting rod for the attraction plate is attached measures 1.4-inches. The distance between the fulcrum and the point on the beam where the connecting rod for the crank is attached measures 2.6-inches. The reduction in motion at the attraction plate end of the beam can be computed as the ratio between two values. See figure 3-11.

$$motion_{plate} = \frac{beam_{plate}}{beam_{crank}}$$

***Figure 3-11**: Relative motion of the attraction plate.*

If we divide 2.6 into 1.4 we get 0.538. This means that, because of the positioning of the fulcrum on the beam, the vertical motion of the attraction plate is only about half the throw of the crank.

Setting conceptual details aside, the physical construction of the walking beam is simple. Starting with a piece of 0.0625-inch (1⁄16-inch) aluminum sheet metal, I cut two strips of metal measuring 4.75 by 0.875-inches wide. These form the sides of the walking beam.

The sides lie parallel to one another, and separated by 5 round, steel threaded standoffs, each measuring 0.625-inch (5⁄8-inch) in length and 0.25-inch (1⁄4-inch) in diameter. The bottom three standoffs function not only as structural members, but also as the attachment points for the connecting rods and the beam's fulcrum.

Obviously I exercised some artistic license in the tapering of the ends of the beam. I also bored large holes in the sides of the beam. I felt that this made the beam more attractive. If these details serve any functional value, it is to reduce the walking beam's overall mass. The components of the beam can be seen in figure 3-12.

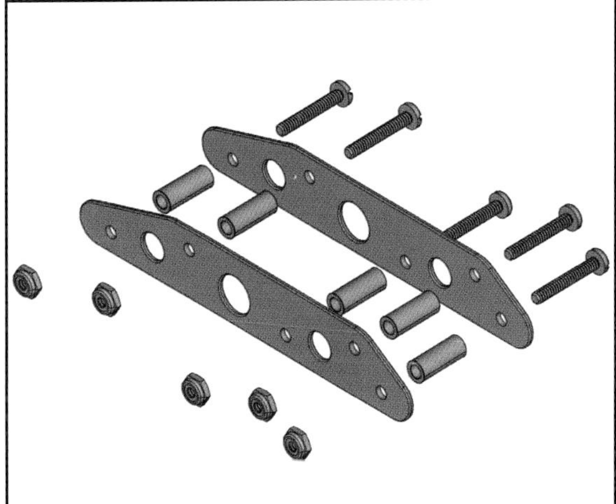

***Figure 3-12**: The beam is comprised of two aluminum side pieces, separated by 1⁄4-inch spacers, and held together with screws and nuts.*

The fulcrum point of the beam rocks to and fro upon a bearing. As I fashioned the bearing and its support from a component extracted from electronic junk, this point warrants a more detailed explanation.

All kinds of junk electronics—including old stereos, amplifiers, televisions, and similar gear—feature potentiometers, more commonly referred to as "volume controls". An example of a potentiometer can be seen in figure 3-13. The potentiometer, or "pot", typically features a squat cylindrical body, a threaded bushing which is used to mount the pot in a control panel, and a rotating shaft which passes through that bushing and into the interior of the pot's body. Inside the pot is a C-shaped resistive element and a metallic contact or wiper that touches the resistive element. The wiper slides up and down the resistor's length depending upon the extent to which the control shaft has been turned. Salvaged pots are often recycled (intact) into my own electronics projects. But even if they are worn and unserviceable, they can be dismantled and used for other purposes.

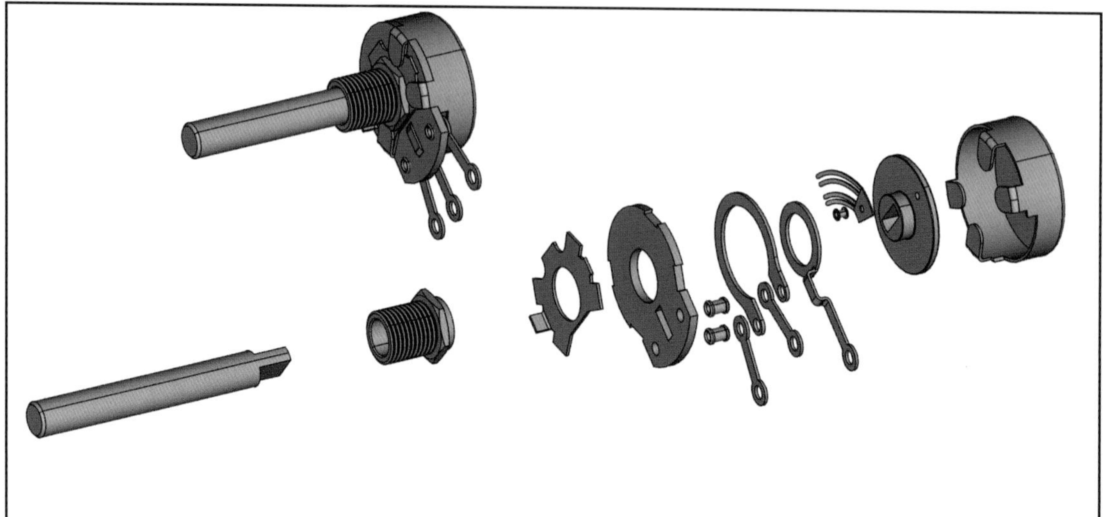

Figure 3-13: A discarded potentiometer contains numerous parts. The potentiometer's threaded shaft bushing can be recycled to make a smooth-running bearing for small machines like motor models.

A bit of creative demolition with a small flat-blade screwdriver, a file, and perhaps a hobby grinding tool, is all it takes to open the pot's body. This allows for the removal of the shaft and the pot's internals, and for the extraction of the bushing. Older American-made pots typically feature ¼-inch shafts, which means that the bushing has a ¼-inch bore. It is one of these pot bushings that I re-purposed as a bearing for the walking beam.

Using scrap aluminum, I fashioned two L-brackets. I bored a hole in each one, large enough to fit the bushing. I installed the bushing on one of the L-brackets, and fastened it with the matching bushing nut. I added a second nut, just to occupy space, and then added the second L-bracket. The latter is held in place with a third nut. The completed assembly can be seen in figure 3-14. Note how the legs of the L-brackets point in opposite directions, providing a very stable base for the bearing and a convenient surface by which to mount the assembly onto another surface.

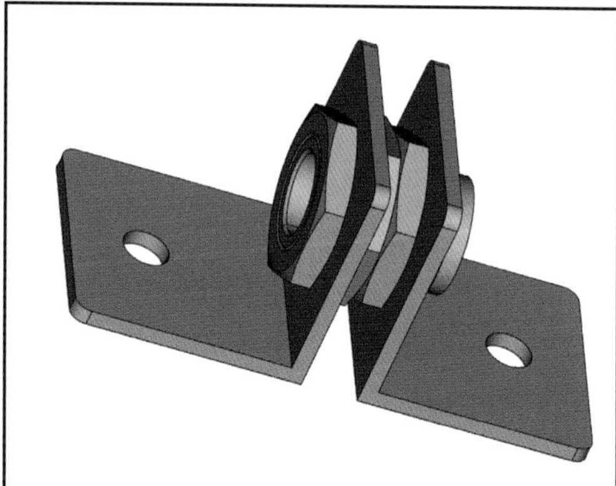

Figure 3-14: The beam bearing assembly is comprised of two L-brackets, a threaded bushing and three nuts.

Recall that the spacers which bond the two halves of the beam together are round, and ¼-inch in diameter. The spacer at the fulcrum point will fit perfectly inside of the bearing I just described and will rotate smoothly. Obviously, one must insert this spacer into the bearing before bolting it into place in the beam assembly.

The walking beam sits about 6-inches above the base of the motor. It's supported in this elevated position atop a "gallows-like" arrangement, that is to say, two legs and a bridge-piece. The legs are aluminum hex standoffs, threaded for #8 screws. Alternately, you could probably use sections of metal tubing or short brass or steel pipe nipples.

The bridge is fashioned from a 3-½-inch section of 0.75-inch (¾-inch) aluminum angle. It provides a shelf upon which the fulcrum bearing assembly is mounted. I drilled holes through the "feet" of the bearing assembly, and corresponding holes in the gallows' bridge. The bearing is attached to the bridge with two 6-32 bolts and nuts. The gallows assembly can be seen in figure 3-15.

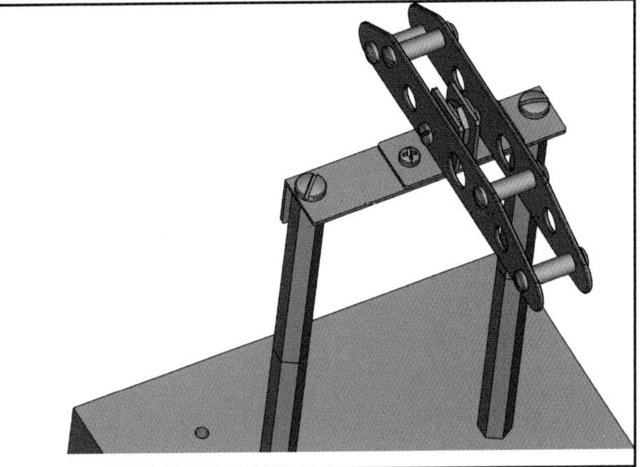

***Figure 3-15**: The beam bearing assembly sits atop a gallows-like structure, comprised of hex standoffs and a bridge. In this image the walking beam is shown installed, as well.*

Connecting Rods

The Texas Motor uses two connecting rods. One is dog-bone-shaped, with a coupler at each end, and is used to link the crank to the walking beam. The other connecting rod is used to link the walking beam to the attraction plate, so it has a coupler at only one end.

The couplers at the ends of the connecting rods were fashioned from shaft collars. The bore of the shaft collars used is 0.25-inch, so they will fit on the crank and the round standoffs at the ends of the walking beam. The hole in the collar where the set screw would normally go was drilled out and re-tapped for a 6-32 thread. Into this hole, I threaded a steel hex standoff. This particular standoff had a 6-32 stud at one end, and a 6-32 threaded hole at the other. One could use a standoff with two female ends, but one end would have to be fitted with a section of threaded rod to act as a stud. Details of the connecting rods can be seen in figure 3-16.

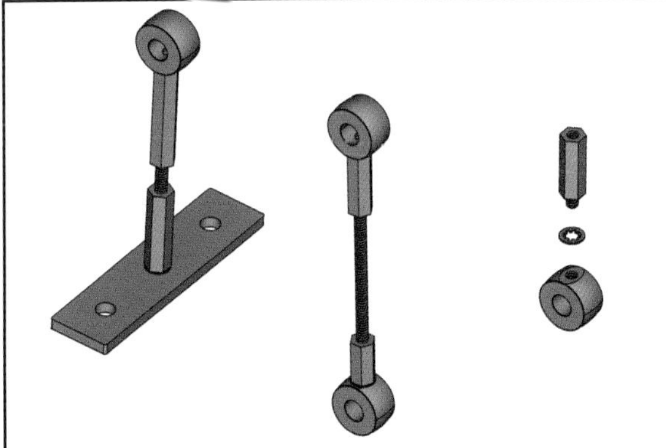

***Figure 3-16**: Connecting rods. At left, the connecting rod for the attraction plate. Center, the connecting rod for the crank. The detail at the right shows how the ends of the connecting rod may be fashioned from a shaft collar and hex standoff.*

The stud on each standoff, even when fully tightened into the shaft collar, does not penetrate into the bore of the collar. I achieved this through some trial-and-error, removing the standoff from the collar, carefully dressing the stud on the standoff with a file, and then reinstalling it to check for interference.

The body of each connecting rod is nothing more than a length of 6-32 threaded steel rod. The couplers are attached to the ends of the rod by screwing them on.

Connecting rods made in this fashion have several desirable attributes. They are cheap and simple to make, they run well with reasonably low friction, and they're compliant. In addition, their length can be adjusted over a limited span by advancing or backing off the standoffs on the threaded rod. This can be very useful when trying to fine-tune the Texas Motor.

Wiring and Engine Timing

As is always the case, the Texas Motor relies on proper timing to establish just when the coils of the electromagnets are to be energized. Like the Peewee, the magnets in the Texas Motor are fired with a leaf switch, and the basic principles with regard to timing apply here, as well. However, there are a few details unique to the Texas Motor.

The first consideration has to do with coil polarity. Recall that the polarity of the field produced from the poles of an electromagnet depends upon the direction in which electrical current circulates through the electromagnet's windings. The pair of coils in the Texas Motor are able to exert the most aggressive force on the attraction plate only if the magnetic polarity of their respective cores is different. In other words, when energized, one core should produce a north magnetic pole, while the other should be a south pole.

With care, I could have recorded the direction with which each coil was wound, noting which lead of the coil fed the center, and which was connected to the outermost winding. With this information, I could have predicted what the pole polarity would be for any given current applied to the windings.

A simpler method is to prepare the coils without regard to polarity, and then characterize them later. This can be done with a flashlight cell and a cheap magnetic compass. To do this, I connected the leads of one of the coils to the flashlight cell and brought the compass near the pole face. The compass needle is either attracted to or repelled by the field produced in the coil. I swapped the leads on the cell, as needed, until I observed that the compass needle was repelled, meaning that the pole of the electromagnet had assumed a north magnetic polarity. When this condition was achieved, I labeled the coil lead which was attached to the positive terminal of the cell with a piece of tape. I repeated this same process with the second coil.

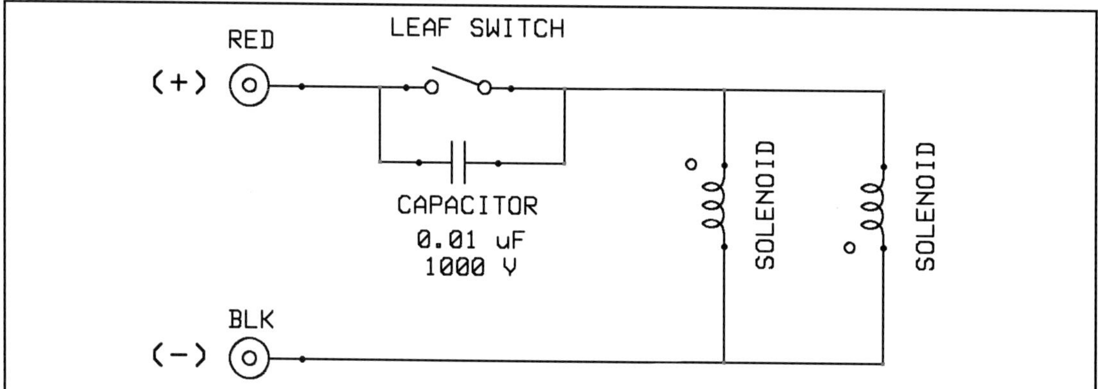

Figure 3-17: *Wiring for the Texas Motor. The dots near the coils signify the coils are wired to produce opposite polarity. A capacitor is used to suppress arcing at the switch contacts. See text for details.*

The coils in the Texas Motor are connected in parallel. However, since we want the magnetic polarity of the two coils to be different, it follows that the electrical connections for one of the two coils must be inverted. This is evident in the wiring diagram in figure 3-17.

The Texas Motor's leaf switch was mounted on the floor of the L-bracket that supports the crankshaft bearing and flywheel. Details of this can be seen in figure 3-18.

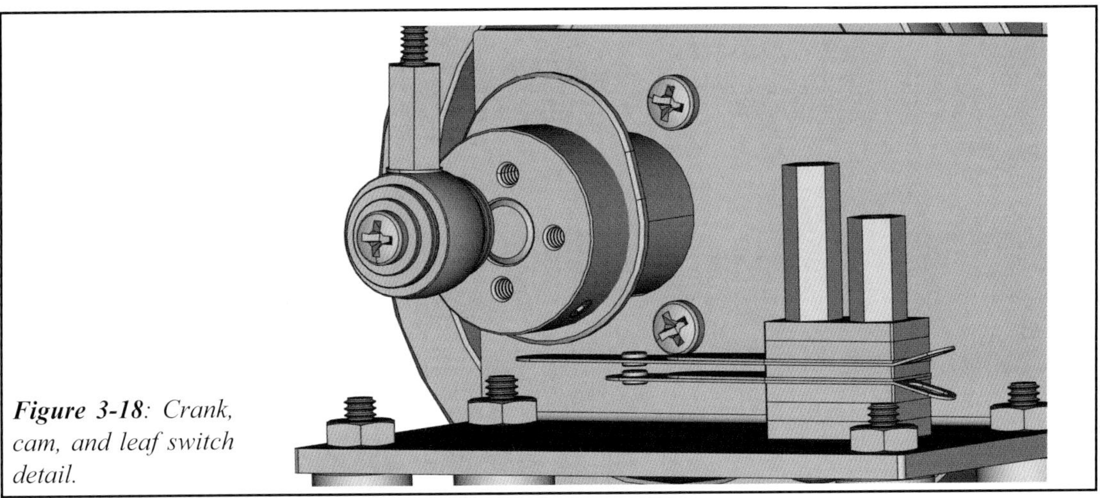

Figure 3-18: *Crank, cam, and leaf switch detail.*

Proper timing for the Texas Motor was established experimentally. The motor was set up to run counter-clockwise, as viewed from the crankshaft end. If we consider top dead center — the position at which the attraction plate is closest to the coils — as zero degrees, the timing cam on the Texas Motor was finally adjusted to fire the coils at around 210 degrees.

Testing and Results

Like the Peewee Motor, the Texas Motor is not self-starting. That said, it exhibits a fair amount of torque for a simple attraction motor. It will sometimes start by itself if the motor happens to have come to rest with the leaf switch in the closed position.

In operation, The Texas Motor has all the charm of the Peewee Motor, and then some. Being more complex, mechanically speaking, and larger to boot, there is simply more to see. It makes a very pleasing sewing-machine-like sound when in operation. This motor seems "happiest" when running at 24V. The stall current (I) for each coil (leaf switch on, but flywheel not turning) is dictated by Ohm's law. See figure 3-19.

$$I = V/R$$

Figure 3-19: *Stall current is computed by using Ohm's law.*

The stall current for each coil is 24V divided by 66Ω, which works out to be 0.36A, plus some change. The stall current for both coils, is 0.73A. That represents a fair amount of power (more than 17W), and is likely to overheat and burn up the coils if left in that state for very long.

Interestingly, when the motor is in operation at 24V, the average current for both coils together is only about 0.12A. This is because current is only drawn when the coils are actually on, and owing to the shape of the cam that operates the leaf switch, the coils are off for most of the flywheel's rotation. The ratio of on-time to off-time, a characteristic engineers refer to as "duty cycle", is therefore quite low. After prolonged operation in a properly adjusted motor, one can detect a slight increase in temperature in the coils, but certainly nothing that would degrade or damage them.

I have run this motor on as little as 10V, where it "chuffs" along, barely turning. I have run it as high as 30V. At 24V, the flywheel turns at a rate of 700 RPM, give or take. The maximum observed speed was just under 1000 RPM. Table 3-20 shows the relationship between motor speed and applied voltage.

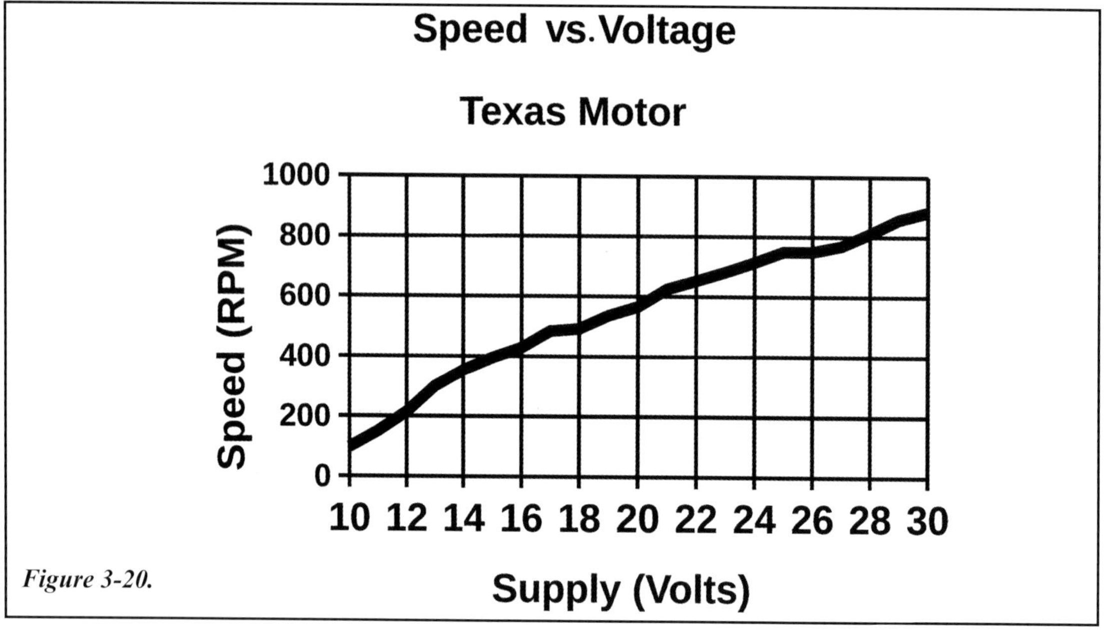

Figure 3-20.

Over the span of voltages tested, this relationship appears more-or-less linear. If I continued to increase the applied voltage, the motor would undoubtedly run faster, though there would certainly come a point at which friction and windage losses would result in diminishing gains. Long before that happened, I would expect problems with heat buildup in the coils and erosion of the leaf switch contacts due to the larger peak currents they would have to carry.

Critical Considerations

All of the criticism voiced in regard to the Peewee Motor also applies to the Texas Motor.

Most of the time, the Texas Motor is not self-starting, and it normally runs in only one direction. To change that direction, one must change the angular relationship between the cam and the crank. And, like the Peewee Motor, the Texas Motor is unbalanced. Vibration is apparent when the machine is in operation, though the sheer mass of the base, the flywheel, and the other components work to suppress it. One could take steps to add a counterweight to the crank or even little weights to the flywheel, but given all of the moving parts, complete balancing seems all but impossible.

The collapse of the magnetic field in the coils when the leaf switch opens produces a hot arc between the switch contacts, just as it does in the Peewee motor. In this motor, however, I took steps to try to suppress that arcing by bridging the contacts of the switch with a capacitor. I describe capacitors and their characteristics in considerable detail in my book *The Voice of the Crystal*, so I won't repeat that material here. Suffice it to say that in this application, the capacitor works against rapid changes in voltage, which can express themselves as arcing between switch contacts. The capacitor I used is a 0.01 microfarad disk capacitor with a 1000-volt rating. This part was harvested out of an old television or computer monitor. Purchased new, it would probably set you back less than fifty cents. With this capacitor in place, the arcing at the switch contacts is still apparent, but is greatly diminished.

Speaking of collapsing fields, the counter-EMF produced by this motor can be substantial. If you throw a neon bulb across the coil terminals or across the switch contacts, you will observe it flash. Touch the switch contacts or wiring while the motor is in operation, and you will probably get poked. Remember—hands off. You shouldn't be sticking your fingers into a machine with moving parts, anyway.

I was very pleased with the performance of my homemade coil bobbins, and expect to make more for future projects. I did, however, want to share a few thoughts on effective coil winding.

Fitting the most windings on any given bobbin depends on your ability to lay down nice, even, layers of wire. Laying down an even layer of wire is critically dependent upon the flatness of the previous layer. In fact, irregularities in any given layer are only amplified as subsequent layers are laid down. All in all, this means that for best results, the center of the bobbin should be a smooth as possible.

The flange "fingers" described earlier represent a somewhat irregular surface on which to lay down the first layer of wire. Therefore, I recommend that after the flanges have been glued into place, you wind and glue a strip of paper inside the bobbin to cover up the fingers, and to assure that the floor of the bobbin is, in fact, perfectly smooth. This should be done before the bobbin is painted and urethaned.

During the winding operation, be very careful if you guide the wire onto the spool with your fingers. Thirty-gauge wire is very thin. If you wind the bobbin too fast, and the wire is drawn between your fingers too quickly, it will cut you as readily and as deeply as any knife. Please exercise extreme caution.

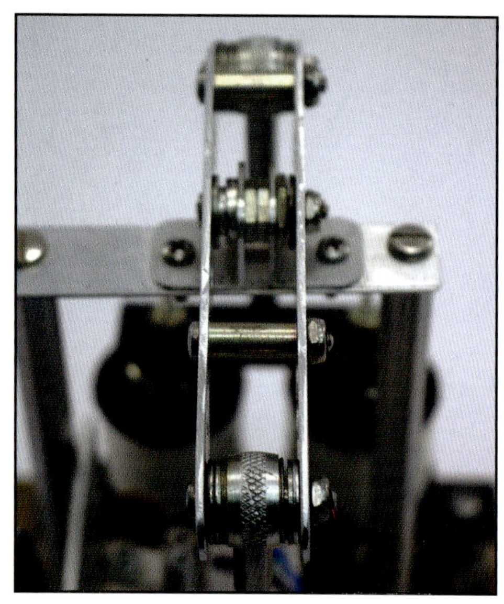

Chapter IV
The Christmas Motor

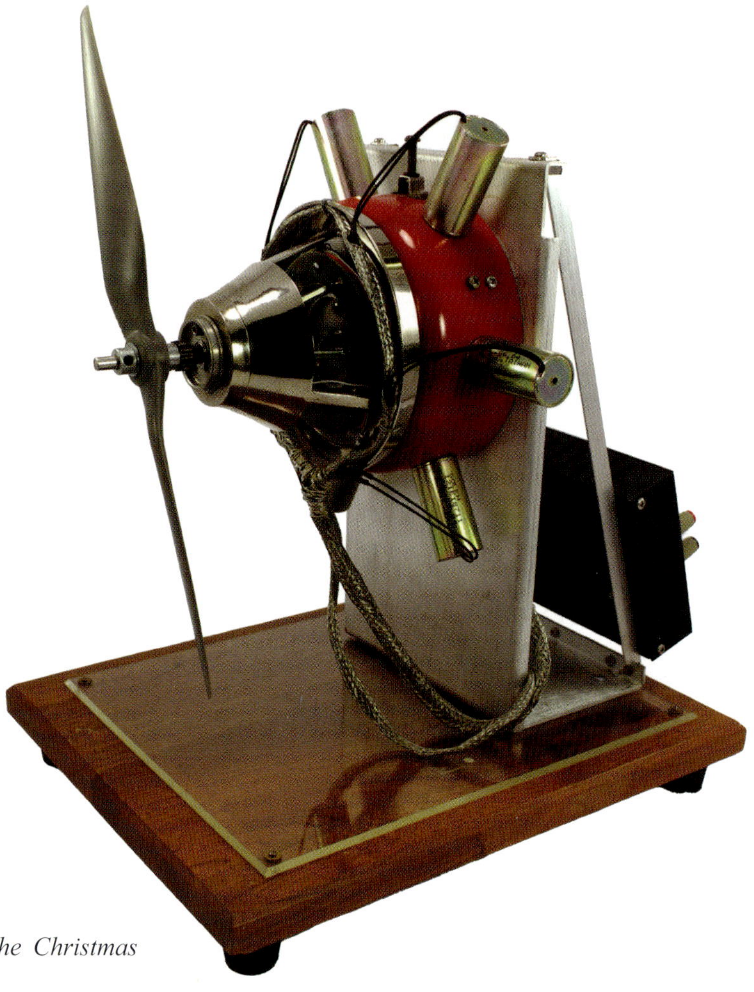

***Figure 4-1**: The Christmas Motor, front.*

One of the recurrent problems with attraction motors, as represented by the Peewee and Texas motors, is the fact that they are inherently non-self-starting. This is easily understood if we consider the operation of the cam and leaf switch in the Texas Motor, for example. Note that the switch is only on for about 60 degrees of any complete rotation of the flywheel.

Suppose I were to give the flywheel of an un-powered Texas motor a random spin, like the proverbial "wheel of fortune". Since the leaf switch is activated only 60 degrees out of a possible 360, the odds of the flywheel coming to rest with the switch in the "on" position is only one in six. More important to the point, for any given throw of the wheel, there is an 83% likelihood that the flywheel will not come to rest in a position where the leaf switch is on. An "off" switch means no power to the coil, and no power to the coil means no starting torque.

Therefore, if we kill the power to a running Texas Motor, let the flywheel come to rest, and then reapply the power, the odds are very much against us with regard to the possibility that the motor will restart by itself. For a model motor, of course, this is no big deal. All you've got to do is give the flywheel a little "kick" to set things into motion again, but for someone with an engineer's mind, this trait is inherently annoying.

Now, we could certainly change the cam profile to lengthen the dwell time of the switch, but the most you can do this is 90 degrees. Beyond that, an "on" coil no longer contributes to torque, but begins to detract from it.

Suppose, however, we expanded the crank, added more solenoids, plungers, and connecting rods, and distributed them so that the "on" periods of their respective coil switches overlapped. In that case, for any given rest position of the flywheel, at least one switch would be on, and at least one solenoid would be positioned to exert torque on the crank when power to the motor was switched on. An attraction motor of this design could, in fact, be self-starting.

By definition, a motor of this type is significantly more complex than the Peewee and Texas motors combined, because of all of the additional moving parts. Then, too, one must give careful consideration to the matter of topology. How many electromagnets will be employed, how will they be laid out, and how can one fashion the complex crankshaft that is needed?

After much thought, I decided to adopt a design reminiscent of old-fashioned aircraft-style radial engines. My design features six solenoids acting on a common crank.

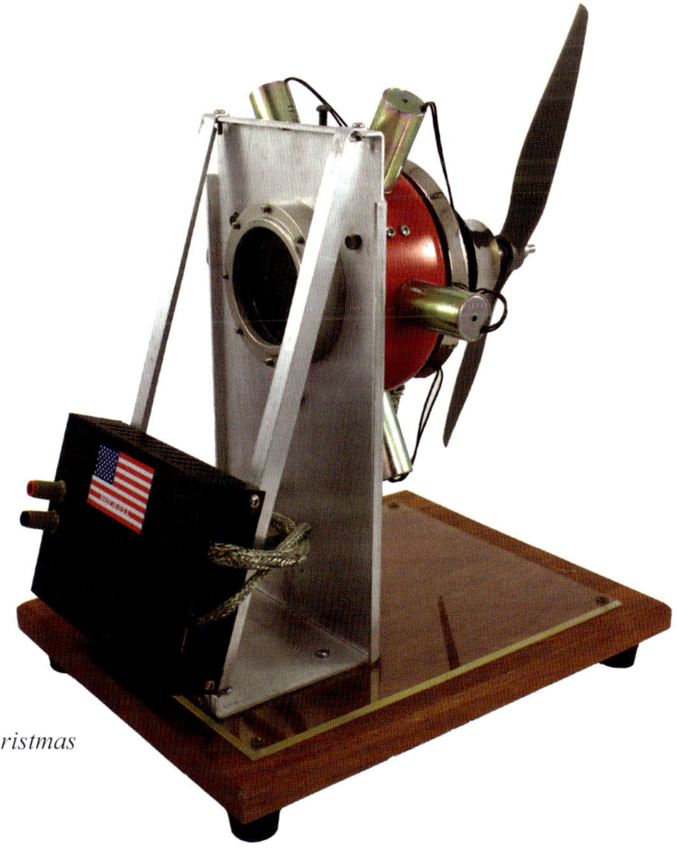

Figure 4-2: *The Christmas Motor, rear.*

I take pride in the design because, despite having seen endless examples of model electric motors in books, magazines, and on the Internet, I have yet to see one built quite like this one.

It took a lot of scrounging in resale shops, garage sales, and garbage cans to find the parts I needed and to conceive of the ways in which the constituent junk could be stitched together. After a lengthy development period, I finally finished the motor a few days before Christmas one year, hence the name, the "Christmas Motor." See figures 4-1 and 4-2.

The Solenoids

Motive force for this motor is provided by a set of six solenoids. In the interest of symmetry and balance, I decided early on that it would be beneficial if I could locate six identical solenoids. This can be difficult if one is relying entirely on cast-offs and junk to provide the necessary parts. Electronics surplus houses should not be overlooked as excellent sources of motor materials. A few minutes on-line located numerous sources of cheap solenoids, including the ones I ultimately ordered for use in the Christmas Motor.

The solenoids I ended up purchasing were, from what I gather, originally manufactured for a well-known maker of printers and photocopiers. They're cylindrical, encased in a metal can or housing, and measure 2-inches long by 1-inch in diameter. The plunger is black steel, just under 2-½-inches in length, and 0.437-inch in diameter. The end is slotted and drilled to allow a connecting link to be secured to the plunger with a drift pin. This coil has a long stroke, possibly as much as an inch. See figure 4-3.

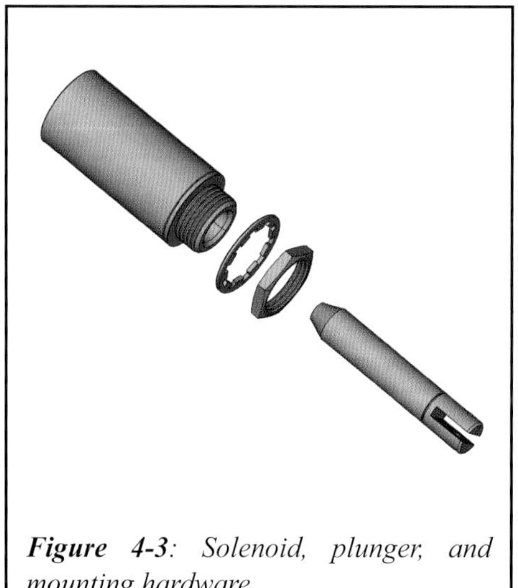

***Figure 4-3**: Solenoid, plunger, and mounting hardware.*

One of the more interesting attributes of this solenoid is its mounting method. Rather than rely on mounting holes for screws, the mouth of the solenoid features a large threaded bushing. The solenoid is designed to be inserted through a hole in a sheet-metal surface, and secured with a lock washer and nut. This is almost made-to-order for the design I had in mind.

The electrical resistance of the coil measures around 45Ω, and the part was designed for 24V service. Yet, in experimenting with the part on the workbench beforehand, I noted that with as little as 6V applied, I could detect meaningful pull on the plunger with it extracted as much as ½-inch.

The specific part I used, for the record, is a Solen 121E18711. They were purchased online for less than $2.00 each, a very small price to pay to acquire a matched set of identical solenoids. The surplus vendor I acquired them from no longer has them in stock, though a subsequent search of the Internet shows that other suppliers do. It's no matter in any event because there are, no doubt, a hundred other surplus coils that could be pressed into service.

The Crankcase

Having decided on a radial-engine design for the Christmas Motor and having acquired a matched set of solenoids, my next action item was to identify something that would service as a suitable crankcase. After visiting all of the second-hand and thrift stores in my area over the course of two or three weekends, I finally happened upon a church rummage sale at which I discovered the perfect component: a one-quart aluminum saucepan. The pan cost me twenty-five cents.

The pan I found is 6-inches in diameter and 2-½-inches deep. The exterior of the pan sports a red coating of some kind, a feature that saved me the trouble of having to paint the motor later. The handle of the pan was broken. This was of no consequence to me, as I simply backed out its screw and threw it away.

I wrapped a narrow strip of paper around the exterior of the pan and then carefully marked the paper with a pencil to record the pan's circumference. Then, with the paper lying flat on my bench, I measured that circumference with a rule, and divided that length into six equal segments. I made pencil marks on the paper to denote these divisions. When I wrapped paper around the pan once more, the marks I had made on the paper indicated precisely how the solenoids should be distributed. I used a permanent ink pen to transfer the marks from the paper to the walls of the pan itself.

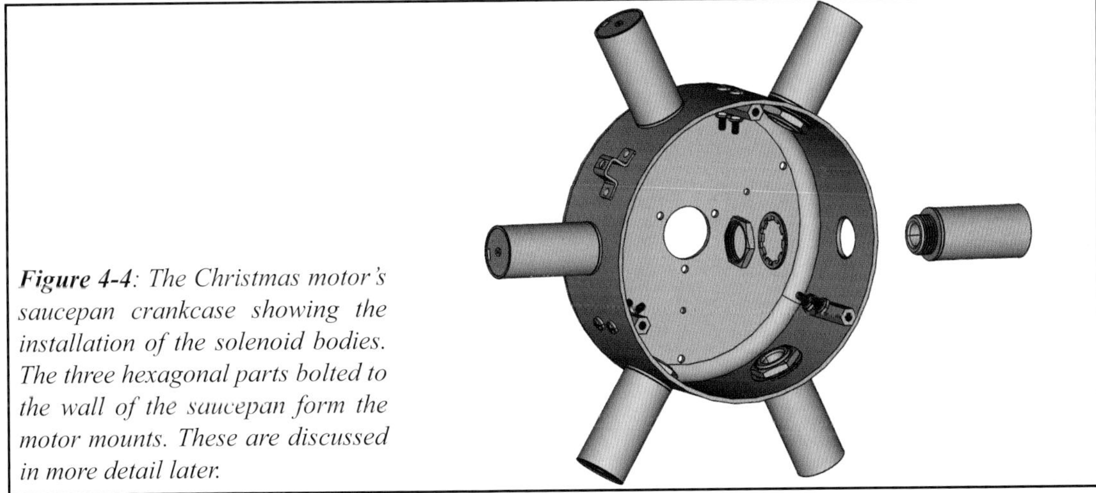

***Figure 4-4**: The Christmas motor's saucepan crankcase showing the installation of the solenoid bodies. The three hexagonal parts bolted to the wall of the saucepan form the motor mounts. These are discussed in more detail later.*

Accounting for the space that the crank would ultimately consume near the floor of the pan, I elected to set the row of solenoids so that their centers lay 1-inch off the open mouth of the pan. I used a rule to make this measurement, and where this 1-inch target depth intersected the pen marks I made earlier, I dimpled the metal with a punch. This activity produced six equally-spaced punch marks denoting where mounting holes would be drilled.

The diameter of the threaded bushing at the mouth of my solenoids was such that each mounting hole had to be ¾-inch in diameter. Aluminum is a very easy-to-work material, so there is more than one way to make holes of this size. My recommendation is to use a step drill bit.

Step bits are cone-shaped drill bits that are capable of making a series of increasing-sized holes, depending on how far the bit is pushed through the material being drilled. High-quality, professional-grade, step drill bits are hideously expensive, perhaps as much as a hundred dollars each. Chinese imports, can cost as little as ten dollars. Now, it's certainly true that you get what you pay for, and cheap Chinese tooling is notorious for its often abysmal quality. However, given the relative softness of aluminum, I've found that even these inferior tools will make nice, clean holes.

Figure 4-4 shows the crankcase with some of the solenoids' bodies installed.

Crankshaft Bearings and the Front Bearing Support

Both the Peewee Motor and Texas Motor made use of the same kind of flange-mounted sleeve bearing for their crankshafts. I could have employed a similar design here, but I like to make best use of the pieces and parts I have on hand. At the time the Christmas motor was being designed, I had just found part of an old computer plotter, which I stripped for parts. Among the treasures extracted from this machine was a section of 0.25-inch steel shaft, and a pair of ball bearings. An example of such a bearing can be seen in figure 4-5.

***Figure 4-5**: An example of a high-quality bearing salvaged from junk. This particular part has ¼-inch bore and it press-fits into a metal flange for mounting.*

***Figure 4-6**: A brass votive candle holder, minus the stubby candle and the glass globe, makes an ideal bearing support for the Christmas motor.*

Among the possible options for shaft bearings in any machine, ball bearings are usually the most expensive. But, with higher price comes high performance, low friction, and exceptional durability. The bearings I found fit into little metal plates which could be used as flanges to mount them to a surface, making them all the more useful.

I drilled a hole in the floor of the crankcase (the sauce pan) and mounted one of the bearings over the hole using some 6-32 screws and nuts. The 0.25-inch shaft I harvested with the bearings passed through the bearing into the interior of the pan. So far, so good.

It was readily evident, however, that a single bearing on the crankshaft could not possibly provide adequate support for the shaft. If any significant radial (side-to-side) load was placed on the shaft, particularly if it extended any distance from the bearing, the effect was to twist the bearing right out of its mounting flange. Clearly, it would be necessary to use a pair of bearings, one in the wall of the crankcase, and one set somewhere farther down the shaft, so as to prevent any yaw of the shaft.

I considered for some time what kind of bracket I might employ to support the second bearing outside of the crankcase. An obvious solution would have been a metal plate, attached to the exterior of the crankcase through the use of threaded standoffs. The plate would have presented a surface parallel to the floor of the crankcase, and a surface in which to mount the second bearing.

As luck would have it, a thrift-store visit uncovered an ideal alternative; a votive candle holder comprised of a decorative glass bowl fitted with an inverted, cone-shaped, brass casting. The casting had a circular rim which engaged the mouth of the bowl, allowing the cone to hang into its interior without falling in. The cone was truncated, that is, its tip was not pointy, but blunted, which providing a flat surface on which a small candle could rest. Thus suspended in the interior of the bowl, a lit candle would illuminate the glass. Figure 4-6 is a conceptual representation of what I'm talking about. In point of fact, the glass bowl was broken, the casting was of little use to anyone but the builder of a radial solenoid engine, and so I walked away with this treasure for some pocket change.

I drilled three holes in the rim of the brass casting, spaced 120 degrees apart, and three corresponding holes in the floor of the crankcase. Using 6-32 screws and nuts, I bolted the two together. Figure 4-7 shows how the crankcase and the front bearing support are bolted together. It also shows the approximate location and manner of attachment for the bearings.

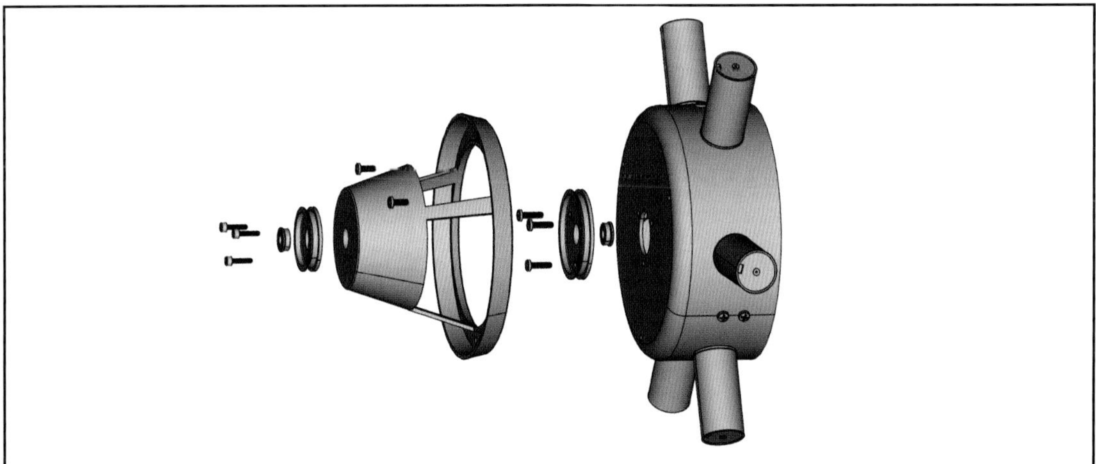

Figure 4-7*: The front bearing support is bolted to the crankcase with three 6-32 screws. Note the placement of bearings and bearing plates. Nuts for the screws are not shown here in the interest of clarity.*

Are you likely to find the same kind of junked computer plotter that I did? Of course not. A logical question, then, is where one might find similar ball bearings. Happily, they're everywhere.

Ball bearings sometimes appear in appliances and tools. Look for them in scrapped drill motors, old blenders or hand mixers, or hobby grinding tools. Old floppy and hard disk drives (the older, the better) are another option. Junked tape recorders, printers, plotters, and photocopiers represent yet other possibilities.

Like solenoids, bearings can be purchased, either brand-new, or surplus. New units are available on the Internet, though bearings comparable to the ones I used run about ten or twelve dollars apiece. Some hardware stores, the kind that have chests of drawers with parts, often stock small bearings, too. At least one hobby shop in my area carries them, for radio-controlled vehicle use, and their prices are significantly cheaper than other sources.

One source of bearings that I find very promising is the local toy store. Skateboards and in-line roller skates use very high-quality ball bearings. Replacements for these bearings are readily available and can be extremely cheap. Although they're typically sold in sets, the per-unit cost can work out to be as little as a dollar apiece.

It is worth noting that the proliferation of inexpensive 3-D printers—whose mechanisms rely heavily upon ball bearings, polished metal shafts, sprockets, belts, shaft collars, couplers and similar machine parts—has driven down the price of these types of components, especially when ordered from sources overseas. Fire up your favorite internet browser and go see what you can find. If you do go this route, be prepared to work with metric dimensions only.

The Crank and Counterweight

The crank design used in the Christmas motor is based on a piece of hardware called a shaft coupler. A shaft coupler is similar in principle to a shaft collar, but it's much longer, and features not one but two set screws, one near each end. Using this device, one can insert a short shaft into each end, tighten their respective set screws, and thereby join two shafts into one, longer shaft.

The shaft coupler used in the Christmas Motor, as most of the pieces and parts in my motors, was salvaged from scrap. It has a ¼-inch bore, and is ½-inch in diameter. They're available for sale new and as surplus, but a part like this is easily made from scratch, using a segment of a ½-inch steel bolt for the body.

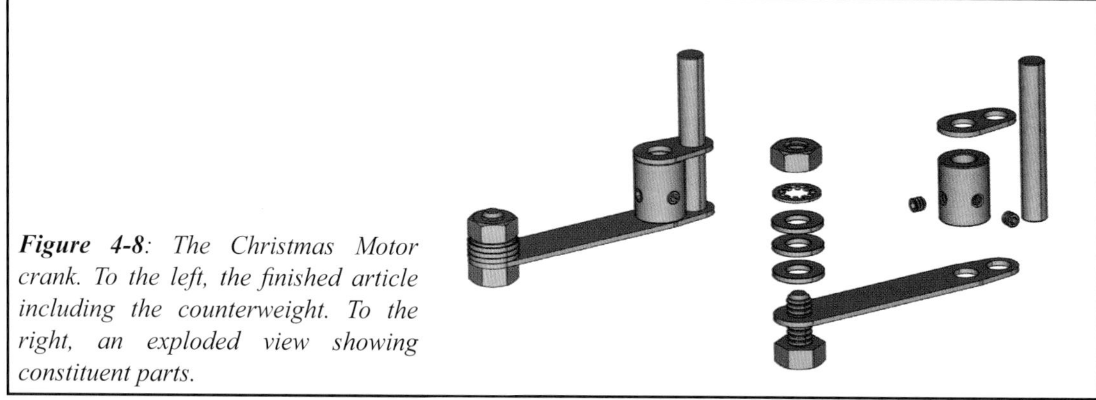

***Figure 4-8**: The Christmas Motor crank. To the left, the finished article including the counterweight. To the right, an exploded view showing constituent parts.*

Figure 4-8 shows how the crank was fashioned. I began with the shaft coupler, and soldered to it a section of ¼-inch rod which forms the handle of the crank. Two end plates, cut and shaped from strips of hobby brass, were used both to add physical strength to the assembly, and to assure that the axis of the coupler and the axis of the crank handle were parallel. To assure proper alignment during fabrication, I temporarily inserted a stainless steel bolt through the end plates and coupler and fitted it with a nut to clamp everything together. This stabilized the parts during the soldering operation.

The bottom plate has an extended arm, about 2-¾-inches long. The end of this appendage provides a mounting point for a small stack of washers that are used as a counterweight—an attempt of mine to provide balance and to suppress vibration when the crank is in motion.

The ½-inch diameter of the shaft coupler, combined with the quarter-inch diameter of the crank handle, results in an overall crank throw of ¾-inch. When coupled to the solenoid plungers through a connecting link, this represents a stroke that is well within the 1-inch capability of the solenoids I used.

The crankshaft itself is a length of ¼-inch steel rod. It fits into the shaft coupler that comprises the body of the crank, and is secured by tightening the set screws in the coupler. For development purposes, I initially used a crankshaft that was far too long. After I had fitted the crank in the crankcase, and allowed for enough shaft length to account for spacers, shaft collars, and the attachment of a propeller to the front of the machine, I marked the shaft, removed it from the machine, and cut it to size with a hacksaw.

The Crank Hub and Connecting Links

The trick, in any radial engine, is to contrive a way in which all the pistons, or in this case, solenoids, can act on a single crank. No part of the Christmas Motor required more thought and planning than this one, and yet, it's ultimately simple in principle and easy to fabricate. Figures 4-9 through 4-16 help to clarify how it was that I fashioned the parts I needed.

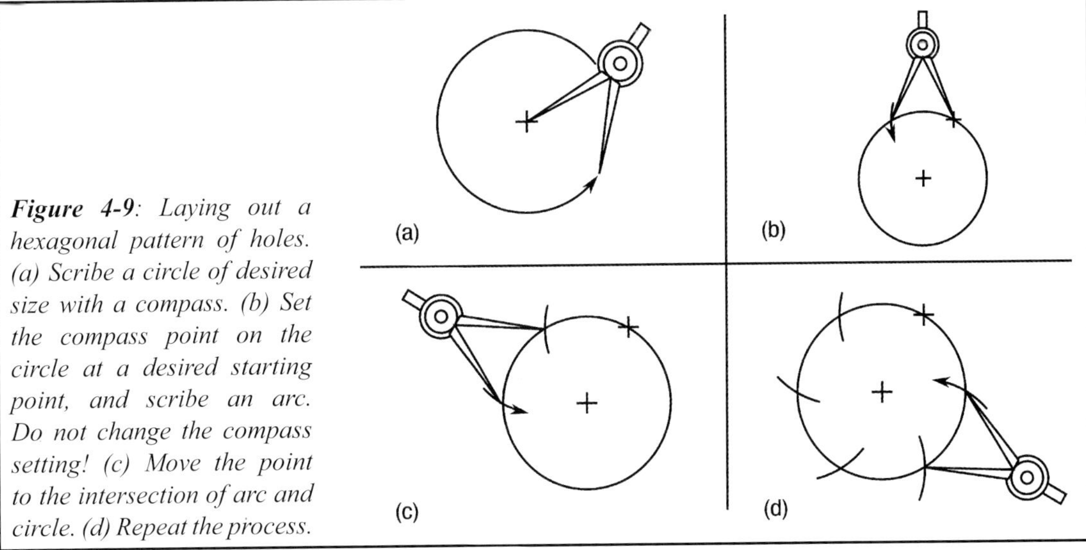

***Figure 4-9**: Laying out a hexagonal pattern of holes. (a) Scribe a circle of desired size with a compass. (b) Set the compass point on the circle at a desired starting point, and scribe an arc. Do not change the compass setting! (c) Move the point to the intersection of arc and circle. (d) Repeat the process.*

Recall that in the Texas Motor, I made use of threaded bushings, extracted from old potentiometers, as bearings for the motor's walking beam. Construction of the crank hub in the Christmas Motor began with one such bushing, along with a collection of three nuts sized to fit the threads on that bushing.

Next, I secured a sheet of 0.062-inch hobby brass and made a shallow dimple in the metal with a punch. This dimple provided a center point against which I could use a compass to scratch several concentric circles into the metal. I set the compass to 0.80-inch, which resulted in a circle with a diameter of 1.60-inches. The purpose of this scratch was to denote the outside border of a disk that I would later cut from that sheet of metal.

I reset the compass to 0.56 inches, and using the same center point, I drew a smaller circle inside the first. This circle was then subdivided into six equal segments using an old trigonometry trick. The key to this trick, which employs the same compass you just used to draw the circle itself, is to leave the compass's span unchanged. In other words, it should be set to the radius of the circle you just drew.

First, you select an arbitrary point on the circle and mark it with a scratch. We'll call this mark "1". You set one point of the compass on "1" and use the other end to strike a new mark elsewhere on the circle, which we'll call "2".

To proceed, you lift the compass, set its point on the new mark "2", and strike yet another mark, "3". Again, you lift the compass, set its point on "3", and strike a mark "4". You continue to walk the compass around the circumference of the circle until six marks have been left. Figure 4-9 shows the process. Done correctly, these marks should be evenly spaced. I used a punch to dimple each of these points.

Finally, I set my compass to 0.43-inch, and drew a third circle in the disk. Using this circle as a guide, I used a punch to create a single dimple which lies between the center of the disk and one of the dimples created earlier.

All of the dimples were drilled out using a 0.062-inch drill bit. The hole at the center of the disk was then enlarged enough to allow it to fit onto the threaded bushing. Finally, with all holes drilled, the disk was cut out of the sheet metal and its edges were dressed with a file.

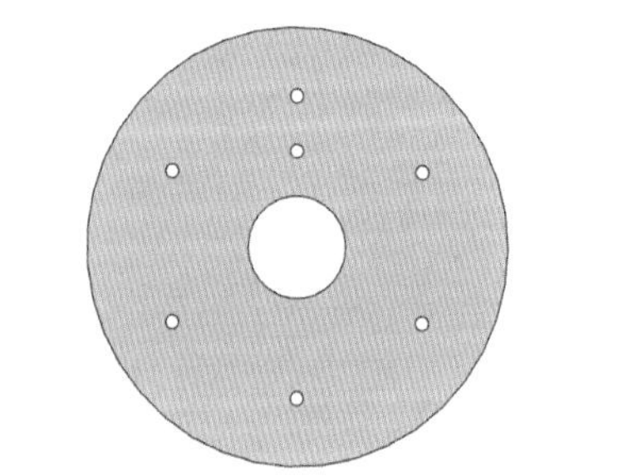

Figure 4-10: *Two brass disks, fashioned as shown, are used to fashion a crank hub assembly. Note the number and position of holes.*

Ultimately, two metal disks, of the type just described, need to be fabricated.

If this verbose description seems confusing, all should become crystal clear with a glance at figure 4-10. There, you see a disk with a large center hole, six tiny holes equally spaced, and a one additional hole near one of the six.

Assembly of the hub begins with the bushing, which is inserted in the center hole of one of the disks, and secured with a nut. Since there is no reason for the bushing and the first disk to ever be separated, I took the added step of heating the assembly with a propane torch to solder the bushing to the disk.

Next, I added the second disk, followed by another nut. Before the second nut was tightened, I rotated the second disk to assure that its holes were in alignment with the holes in the first disk. See figure 4-11. This completes the basic assembly of the hub. There are a few more parts that go with it, but it's best to diverge for just a moment to describe the links that couple the solenoid plungers to the hub.

The links I used were fabricated from hobby brass strips, ½-inch wide, 1.375-inches long, and something on the order of 0.062-inch thick. Near one end, the end that was to couple with the crank hub, I drilled a 0.062-inch hole. At the other end, the end that was to couple with the solenoid plunger, I drilled a 0.125-inch hole. Needless to say, six connecting links must be fabricated, five of a simple type, and a single, slightly different, "master" link. Why the master link differs from the other links will be addressed in just a moment. See figure 4-12.

Figure 4-11: *The basic crank hub assembly is comprised of two brass disks, two nuts, and a threaded bushing. On the left, the constituent parts; on the right, the assembled hub.*

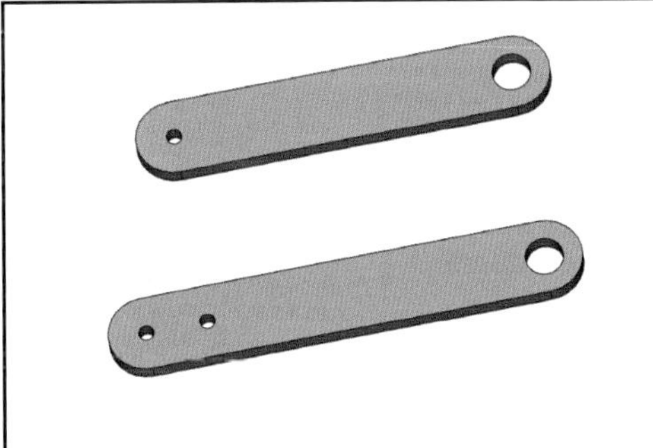

Figure 4-12: *Examples of connecting links. The link depicted at the top is replicated five times. The link at the bottom is a so-called "master" link. Only one of these is needed.*

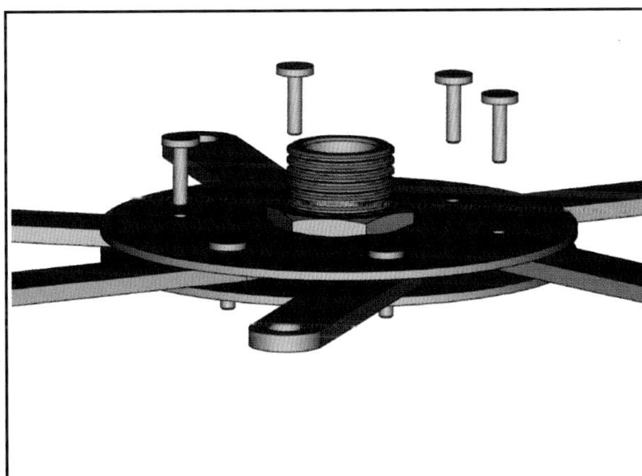

Figure 4-13: *Connecting links swivel on and are pinned to the hub with modified box nails.*

Now this is where things start to get interesting. To couple a given link to the hub, I inserted the link in between the two disks on the hub, and maneuvered it about until the hole in the link aligned with one of the six holes in the disk. The link was then pinned into place by inserting a finishing nail through the hole in one disk, through the link, and through the corresponding hole in the second disk. Again, this is a very verbose way of saying that the two disks on the crank hub act like a clevis. The insertion of a steel box-nail pins the link into place. This can be seen in the edge-view in figure 4-13.

The pinning process was repeated until all six connecting links had been pinned to the crank hub. Note that it may be necessary to trim the link pins for length. If they protrude too far, there is always the danger that they may interfere with the motion of the crank, which operates in the space below.

Now, if you've been paying attention, you must surely wonder what keeps the link pins from falling out. Good question. The answer lies in the fabrication of a third disk, a retaining disk. This has the same diameter as the first two, with a single large hole at the enter, but without all of the other little holes. In essence, it's nothing more than a fender washer. I placed it on the bushing, and secured it with the last nut. With this retainer in place, the pins can jiggle and even back out of their holes a little bit, but they can't fall out because the disk keeps them from backing out all the way.

For what it's worth, I took the added step of cutting out a cardboard ring or "doughnut" to fit underneath the retainer disk. It takes up all of the available "slack" between the heads of the link pins (finishing nails) and the retainer disk, eliminating in-and-out jiggle in the link pins entirely. Figure 4-14 is a more elaborate exploded view of the hub assembly, showing the addition of the pin retaining plate and cardboard ring. A final lockwasher and nut holds the whole thing together.

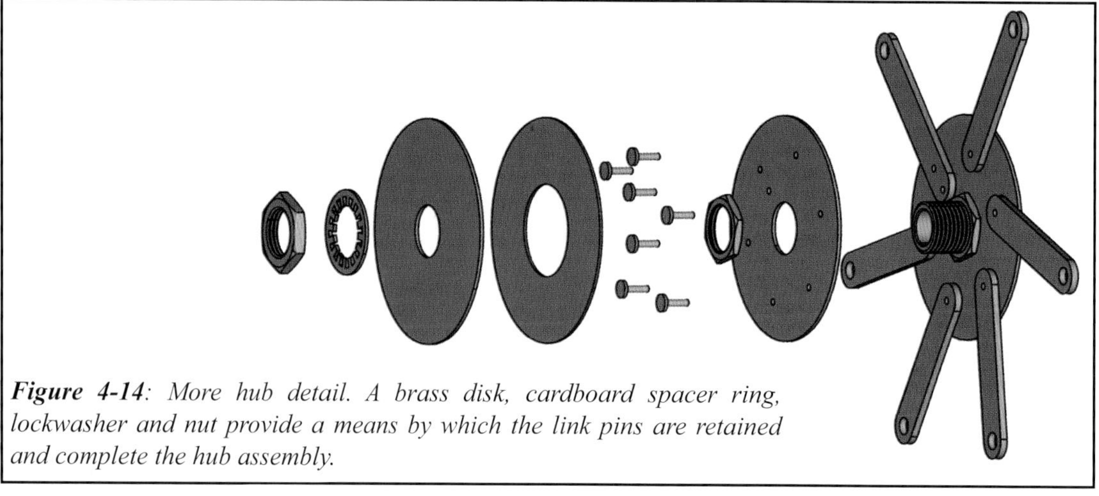

***Figure 4-14**: More hub detail. A brass disk, cardboard spacer ring, lockwasher and nut provide a means by which the link pins are retained and complete the hub assembly.*

With the hub completely assembled and the connecting links in place, I slid the crank hub onto the end of the crank. To keep the hub from falling off of the crank handle, I installed a ¼-inch shaft collar.

The free ends of each of the connecting links were inserted into the clevis (slot) at the end of each solenoid plunger, and pinned into place with a cotter pin. I have found that if the clevis at the end of the plunger is much wider than the thickness of the connecting link, you may experience some mechanical slop as the link slides to and fro. In this case, it's prudent to add some washers to take up the slack. All of this can be seen in figure 4-15.

I have described the construction of the hub and the attachment of the connecting links to reflect a progression of thought. There is, however, an essential detail that I omitted and must now bring to your attention in order for this machinery to work properly.

If all of the connecting links are free to swivel at their point of attachment to the crank hub, there exists the probability that, when forces are applied to the hub by the solenoids, instead of moving the crank, the hub will simply rotate and "wind up" around the crank handle. This condition is prevented, in every radial engine I've seen, by assigning one of the connecting links the role of "master" link. As we've seen, the master link in this engine is different from the other five links in that it is somewhat longer and features two holes at the hub end.

Hopefully this revelation sheds some light on the purpose of that "extra" hole in the hub disks. The master, when pinned to the hub, actually receives two pins, so that it's effectively locked to the hub and cannot swivel. The rigid connection between hub and master link prevents undesired rotation of the hub. Take a closer look at figure 4-14, and look for the double-pinned master link.

With the parts fabricated and installed as described, including the double-pinned master link, I could twirl the crank shaft with my fingers and the plungers would dance inside the crankcase.

The Base, Motor Stand, and Motor Mounts

Radial motor models, particularly when fitted with propellers, look like they belong on the nose of an airplane. As I had no intention of building a matching airframe, I fashioned a motor stand to support the Christmas Motor for display and operation. The motor stand is an L-shaped affair made from a 0.125-inch (⅛-inch) scrap aluminum panel.

I used a machinist's square and an ink pen to mark a line where I intended to bend the panel. I clamped a hardwood block on either side of the panel, adjacent to the ink line, and loaded the whole affair into my bench visc. As I applied force to the free end of the panel with my hands, the hardwood blocks acted like a sheet metal brake and allowed me to make a clean, 90-degree bend.

The motor stand is 12-inches high, 6-½-inches wide, and has a foot that's 4-½-inches long. Because the crankshaft mechanism in this motor is a real thing of beauty, it seemed a shame to me that it would be hidden inside of the crankcase. For this reason, I cut a 2-½-inch diameter hole in the face of the motor stand, and fitted to it the metal bezel and glass from a discarded air pressure gauge.

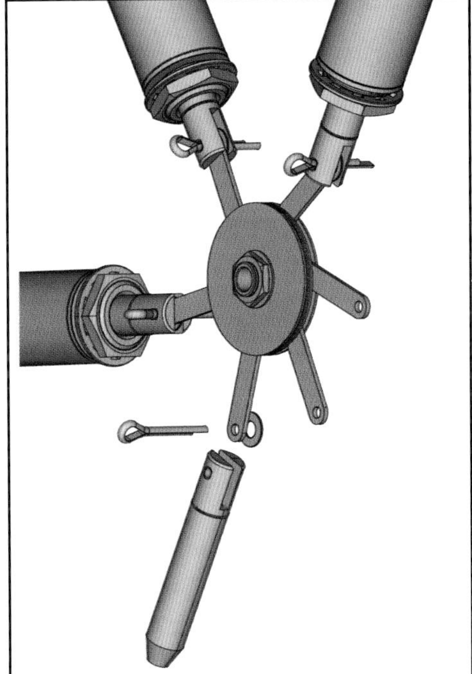

Figure 4-15: *Plungers are secured to the connecting links with cotter pins. Washers can be used to take up the "slop" if the gap in the tail of the plunger is too wide. The solenoid plungers ride in the bore of the solenoid bodies. While this image depicts the mechanism without the crankcase for the sake of clarity, the actual assembly of these components must take place inside of the crankcase.*

This creates a window through which the complex motion of the crank, hub, and connecting links can be observed when the motor is in operation.

Finally, I stiffened up the motor stand by adding a pair of aluminum struts, made from lengths of ½-inch aluminum angle. These tie the top corners of the motor stand to the foot. The result is a triangle-shaped structure of extreme strength and rigidity.

The motor stand is bolted to a finished wooden base. The base itself is actually a re-purposed awards plaque, found at a rummage sale. Using sheet metal screws, I attached four rubber stoppers to the bottom of the base to act as feet. Despite the fact that the Christmas Motor vibrates somewhat when in operation, the compliance of the rubber helps to isolate that vibration from whatever surface the model might be sitting on. The base, struts, and observation window are apparent in figures 4-1 and 4-2.

***Figure 4-16**: Interior of the crankcase showing two of the three motor mounts. The mounts are fashioned from threaded aluminum hex standoffs, each fastened to the wall of the crankcase with two screws.*

To mount the crankcase to the motor stand, I fashioned motor mounts from three threaded, aluminum, hex standoffs. Two holes were drilled through the side of each standoff, comparable holes were drilled in the wall of the crankcase, and the standoffs were bolted to the crankcase with 6-32 screws and nuts. Three holes were drilled in the face of the motor stand, corresponding to the location of the motor mounts, and the motor was bolted to the stand using #8 screws. Two of the motor mounts, installed in the crankcase, can be seen in figure 4-16.

Timing

Solenoids, connecting links, and a crank make for the bulk of what constitutes this motor, but as we saw with the Peewee and Texas motors, there is the matter of solenoid timing. This motor is significantly more complex than the first two, because of the need to activate and deactivate not one, but six solenoids.

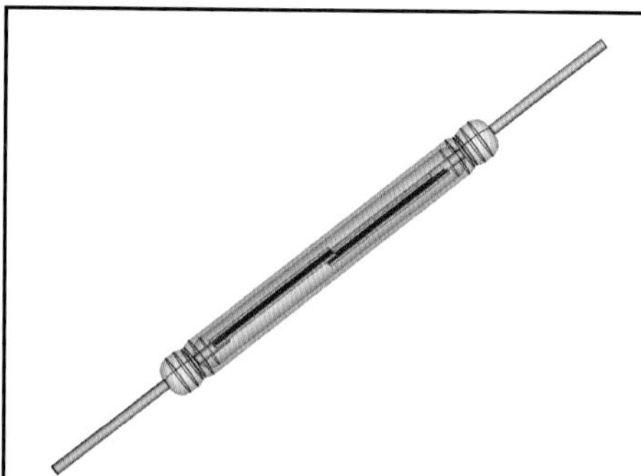

***Figure 4-17**: The reed switch is comprised of two metallic "reeds," separated by a small air gap and sealed in a glass tube. The field from a nearby magnet will cause the reeds to flex and stick together, thereby closing the electric circuit screws.*

There is no reason why a cam and switch arrangement, like the ones used in the first two motors, couldn't be adapted to the Christmas motor. The problem with doing so is primarily one of practicality, because of the difficulty in mounting six separate leaf switches in close proximity to the crankshaft. There are ways to do it, but most of these ways are ugly.

The Christmas Motor makes use of a very special kind of switch called a reed switch. Reed switches are comprised of two metal strips, or reeds, which are sealed inside of a long glass, tube. See figure 4-17. The reeds are set into the glass such that their tips overlap, but do not touch. They are separated by a very narrow gap. When we connect a reed switch in series with a small battery and a lamp, nothing will happen, because the parted reeds present an open circuit.

Magic happens, however, when we bring a magnet close to the glass. The field from the magnet induces a field in the reeds. They are attracted to one another and will stick together, closing the electric circuit. The lamp will light. If we remove the external magnet, the reeds part once more, and the lamp goes out.

The timing concept behind the Christmas Motor is based on a permanent magnet, glued to a metal bracket, which is itself attached to the crankshaft of the motor. Surrounding the shaft, the bracket, and its magnet, we arrange a set of six reed switches. As the crankshaft turns, the bracket whirls about. When the magnet is in proximity to one of the reed switches, that switch is activated. When the magnet passes, that reed is deactivated, and the next reed turns on. Together, one could envision the function of these parts as being reminiscent of the distributor in a six-cylinder automobile engine.

Reed switches are not necessarily common, but they aren't rare, either. They can be found in old alarm equipment, and are used to detect the opening of doors and windows. They are also used, on occasion, in appliances and office machinery to detect the opening and closing of an access panel or door. Barring a junk source, they are readily available in the surplus market, often for less than a dollar apiece. Even brand new, they are cheap and easy to find. In my case, my reed switches were used and were harvested from some scrapped signal switching gear.

As mentioned some paragraphs back, the front bearing support, made of a votive candle holder, is roughly cone-shaped. I conceived that the interior of this cone would be the ideal place to locate my timing components. I reasoned that they would be protected there, and by mounting them within the cone, the motor would retain a nice, compact, and aesthetically pleasing shape.

Starting with some fiberglass circuit board material, 0.065-inch thick, I cut two insulating rings, a large one and a small one. Into each ring I drilled six, equally-spaced holes, large enough to accept the leads of my reed switches. By inserting the leads of the reed switches into the holes in the rings, I created a conical assembly best described as a small ring supported by six reed switches over a large ring. The size of the rings and the resulting slant of the reed switches was engineered to allow the completed assembly to fit just inside of the cone-shaped front bearing support. Figure 4-18 clarifies this idea.

***Figure 4-18**: Reed switches are assembled into a cone-shaped structure, with an insulating ring at the top and at the bottom. Three hex stand-offs provide legs by which the reed cone can be attached to the face of the crankcase.*

The circuit board material I used had copper pads on it, which allowed me to solder the leads of the reed switches to the rings. This bonded the entire assembly into a strong and rigid structure. Alternatives to circuit board material might be plastics like acrylic or polycarbonate. Rings cut from old CD-ROMs might work just fine. Another option might be rings cut from chunks of kitchen counter-top laminate material. Since these alternative materials don't have pads that the reed switches can be soldered to, I would advise gluing the assembly together using a drop of epoxy at every junction between reed switch and ring.

At the narrow end of the reed-switch cone, I used a piece of wire and a soldering iron to join all of reed switch terminals together. I terminated this convergence of switch terminals with a single piece of wire, a couple of feet long. At the wide end of the cone, I soldered an individual wire to each of the reed switch terminals. The end result is a bundle of seven wires, one for each reed switch and one for the common terminal at the narrow end of the cone. I'll discuss the actual circuitry for this motor in just a moment.

Wire can be purchased, but I like to used recycled materials for my motors. In this case, the wires soldered to the reed switches were harvested from an old computer parallel-printer cable. The conductors in these cables are a fairly light gauge, but more than up to the task they're being asked to perform here. One advantage of recycling such cabling is that the conductors inside are often multicolored. By using a different colored wire for each reed switch, the additional wiring that must be done later is greatly simplified, because it's easy to tell at a glance which wire corresponds to which reed switch.

Figure 4-19: *A strip of hobby brass, a couple of magnets, and a shaft collar create a commutator arm. It is mounted on the motor shaft beneath the reed cone. It establishes the timing of the motor by activating the reed switches in rapid succession.*

Aircraft propulsion systems often make use of braided metal hoses for their external plumbing. I like this look and decided to apply it to the Christmas Motor. Again, useful materials were found in computer cables. Some printer and all monitor cables are shielded, meaning that the bundle of conductors at the center is wrapped in a metal-fabric sleeve, woven from strands of wire thread. I identified an old parallel printer cable with such shielding, cut off the ends with wire cutters, and then used a pocket knife to carefully slit and strip off the outer plastic insulation. This revealed the shielding, which bore more than a passing similarity to aircraft hose braiding.

By design, many of these cables have far more conductors inside than we need for the Christmas motor. Using needle-nose pliers, I pulled the wires out of the bundle that I didn't want.

The base of the reed switch cone (the large ring) was drilled with three holes, and using short 6-32 hex standoffs as feet, I mounted the reed cone to the bottom of the saucepan crankcase. Again, it was positioned to fit inside the cone-shaped front bearing support.

Next, I fashioned what I call a "commutator arm". This is a bracket which brings a pair of magnets into proximity of the reed switches as the motor shaft turns. I fashioned it from a length of ½-inch hobby brass strip. I drilled a ¼-inch hole near each end, and then bent the bracket into a shape that, for lack of a better description, is L-shaped. The angle between the foot and spine of the L is acute, meaning less than 90 degrees. The top of the L is folded over, so that the plane of the hole at the top of the L is parallel to the plane of the hole in the tip of the foot. I lined up a ¼-inch shaft collar with the hole in the bracket's foot, and soldered it into place. A picture is worth a thousand words, they say. Figure 4-19 makes a strong case for that.

The bracket is installed on the crankshaft by inserting the shaft through the shaft collar and hole in the foot, and then through the hole at the top of the L. The bracket is locked onto the crankshaft by tightening the setscrew in the shaft collar, and the result is a triangular brass arm that projects from the side of the crankshaft and can whirl about.

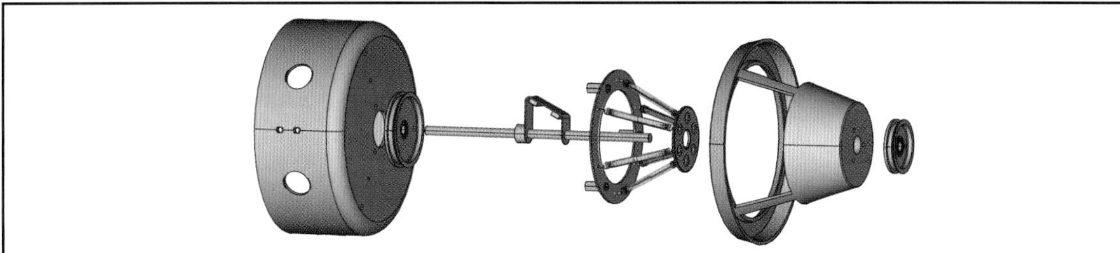

***Figure 4-20**: An exploded view of the relationship between front-end components. The reed cone is nested inside of the front bearing support (which is also conical) and both are attached to the face of the crankcase. The commutator arm with its magnets whirls inside of the reed cone. Also shown here is the motor shaft and bearing plates with their bearings.*

The dimensions of the bracket were engineered so that its form could revolve within the interior of the reed switch cone I described earlier. The foot of the L, then, is sized to be somewhat smaller than the radius of the crankcase-side end of the reed cone, and the spine of the L is sized to slope at an angle that matches the interior angle of the reed cone. The relationship between the reed cone, the commutator arm, and other front-end motor components can be better understood by studying figure 4-20.

The only thing missing from this description is the permanent magnet. I used more than one. The magnets I employed were harvested from the earpieces of a cheap set of MP3 player headphones. Small, yet extremely powerful, magnets can be harvested from the optical head assembly of discarded CD-ROM players, as well as from the worn-out brush-heads of some brands of electric toothbrushes. Small loudspeaker magnets, and even refrigerator magnets may work just as well.

Rather than attach the magnets to the exterior surface of the L-shaped bracket, I epoxied them to the interior. Whirling a magnet around at high speed can generate significant centrifugal forces, and I didn't care to risk having a magnet break free and crash through the fragile glass reed switches surrounding it.

I would also add that some experimentation is warranted in the selection of the number and placement of magnets. As I indicated earlier, I used two magnets. One was placed near the top of the L, and one closer to the foot.

When using multiple magnets, an important factor to consider is their polarity. In my case, I arranged my magnets so that one magnet presented a north magnetic pole to the reed switch, the other magnet presented a south pole. I found that this resulted in greater reed switch sensitivity and more aggressive switch closure.

Remember that where magnets are concerned, like poles repel, while unalike poles attract. Knowing this, and with a moment of idle play, you can easily determine how to orient the magnets on the bracket so that their poles will differ. Alternately, you can use a magnetic compass to positively identify the polarity of any magnet pole. The pole of the magnet that deflects the north-seeking end of the compass needle must, by definition, also be a north seeking magnetic pole.

Needless to say, the angular position of the L-shaped magnet holder on the crankshaft is not arbitrary. It must be adjusted to trigger the firing of each solenoid at the proper time. Let your experience with the Pee Wee and Texas motors be your guide.

Coil Drivers

One might assume that, having reached this point, the description of the Christmas Motor must be nearly complete. After all, we have a bank of solenoids mechanically linked to a common crankshaft, and a set of switches with which to activate them. Surely, now, it's just a matter of wiring each reed switch to its respective solenoid and adjusting the rotating magnet's position to set the appropriate timing. Almost true... but not quite.

We've seen in prior chapters that solenoid coils can draw a fair amount of current. At a DC resistance of 45Ω, the solenoids used in this motor, when powered from a 24V supply, will draw more than 0.5A. Based on the operation of the Peewee and Texas motors, we also know that the collapse of the magnetic field in a solenoid is prone to cause arcing at the switch contacts.

Reed switches are delicate devices. While there are exceptions, most reed switches are simply not robust enough to tolerate this kind of abuse. Exceed their current rating, or subject them to any significant arcing, and the internal contacts are rapidly destroyed.

What is needed is an interface of sorts, something that allows the reed switch to control the solenoid without having to drive it directly.

One way to implement this idea is with a relay. A conceptual example of the circuit can be seen in figure 4-21. The relay here consists of a small coil associated with a robust set of switch contacts. When the reed switch closes, battery power is applied to the relay coil. The relay coil generates a magnetic field, which attracts an iron armature. The movement of the armature closes the relay's switch contacts. When these contacts close, power is applied to the solenoid.

The reason this scheme is advantageous is because the relay coil requires far less energy to fire than the solenoid. Adding the relay limits the electrical burden imposed upon the reed switch. The reed switch tells the relay when to close, but the relay does all the hard work.

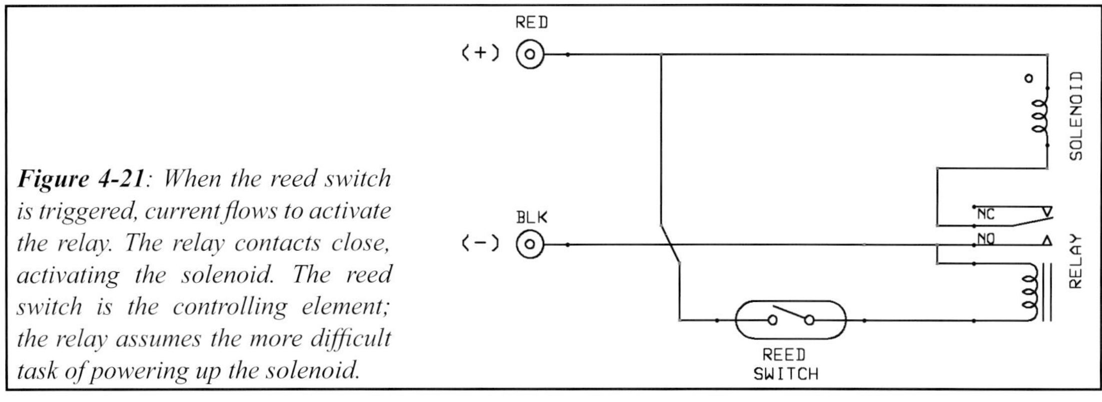

***Figure 4-21**: When the reed switch is triggered, current flows to activate the relay. The relay contacts close, activating the solenoid. The reed switch is the controlling element; the relay assumes the more difficult task of powering up the solenoid.*

A big disadvantage of the relay is its electromechanical nature. It contains moving parts which click and clatter every time the relay is engaged and released. Ultimately, it wears out. Even if a relay is good to a half-million cycles, a solenoid motor running at 1000 RPM will consume that relay's advertised life in just over eight hours.

Reliability can be greatly improved if we replace the relay with a transistor. The transistor is an electrical component that acts as an amplifier. Wired correctly, it can be made to function as a relay. It has no moving parts, so as long as its electrical specifications are not exceeded, there is no practical limit on its lifespan. Figure 4-22 depicts the relay circuit of figure 4-21 revised to use a transistor. Again, this specific design is more conceptual than practical. You'll see why in just a moment.

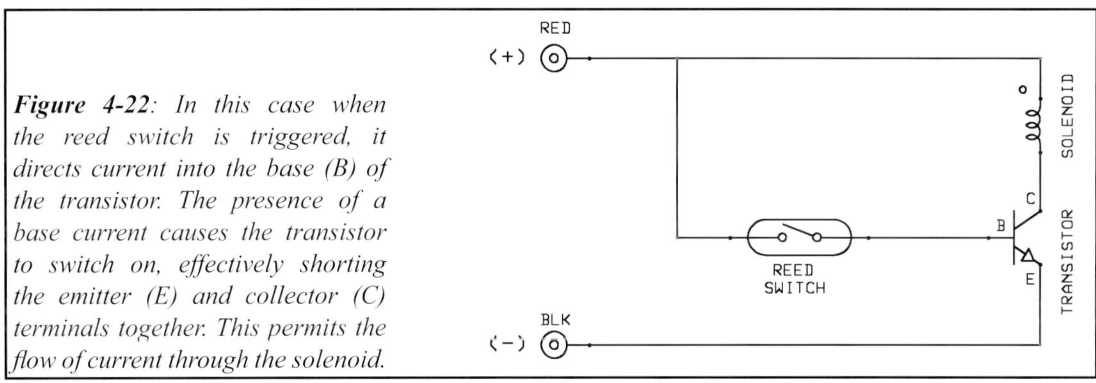

***Figure 4-22**: In this case when the reed switch is triggered, it directs current into the base (B) of the transistor. The presence of a base current causes the transistor to switch on, effectively shorting the emitter (E) and collector (C) terminals together. This permits the flow of current through the solenoid.*

The transistor has three terminals, an emitter, a base, and a collector. I've used the letters E, B, and C to identify the terminals in the diagram. In this application, the emitter and collector can be considered as a set of contact points which are used to energize the solenoid. The base terminal is the input, and is connected to the reed switch. When the reed switch closes, current flows into the base, which activates the transistor. This causes the collector and emitter to be electronically joined together, allowing current to flow through the solenoid. The amplifying capability of the transistor means that a very small current flowing into the base, perhaps 10 milliamperes, can trigger a current in the solenoid that's 50 or more times greater.

Roughly speaking, there are two classes of transistor, designated by the letters NPN and PNP. This nomenclature tells an engineer how voltages should be applied to a given transistor to make it function properly. It's an important matter, because in many cases, applying a voltage of the wrong polarity to a transistor terminal can result in instant and permanent damage.

The transistors used in the Christmas Motor are part number 2N3055. This is an NPN power transistor intended for applications where large currents are being controlled. It's an extremely rugged, common, and inexpensive part, available for more than 50 years. My transistors were harvested from some dead power supplies rescued from a dumpster. They are also available through surplus dealers, dirt-cheap. Other transistor part numbers may work just as well, and can be found in discarded electronic equipment like old stereo equipment and computer monitors. A great source of transistors, for example, is a discarded computer. During the development of the Christmas Motor, I verified that the motor will run just fine under the control of scrap NPN power transistors harvested from dead PC (personal computer) power supplies.

When considering a set a power transistors for a motor like this, one needs to understand the innate limitations of the parts. The electrical characteristics of a given transistor are published by the manufacturer in the form of a data sheet. Usually, data sheets are easy to find and readily available on the Internet. Some of the numbers on a data sheet have little bearing on an application like this, while others are absolutely critical.

For example, a common characteristic found on transistor data sheets is the maximum collector current, usually designated by the symbol Ic. Since the transistor is being used as a relay, all of the current flowing through the solenoid must also flow through the collector. Thus, the transistor's rating for maximum collector current should be larger than the current that the solenoid is expected to draw. The data sheet for the 2N3055 indicates that the maximum collector current for that part number is 15A, a value thirty times greater than the current in the associated solenoid. No problem there!

Another important parameter is Vceo. This is the maximum voltage that can be applied between the emitter and the collector. In a solenoid driver application like this one, the voltage applied across the emitter and collector is equal to the supply voltage for the motor, so ideally, the Vceo value should be larger than the motor supply. In this case, the intended operating voltage for the Christmas Motor is 24 VDC. The data sheet for the 2N3055 specifies Vceo at 60 VDC. Once more, this looks good.

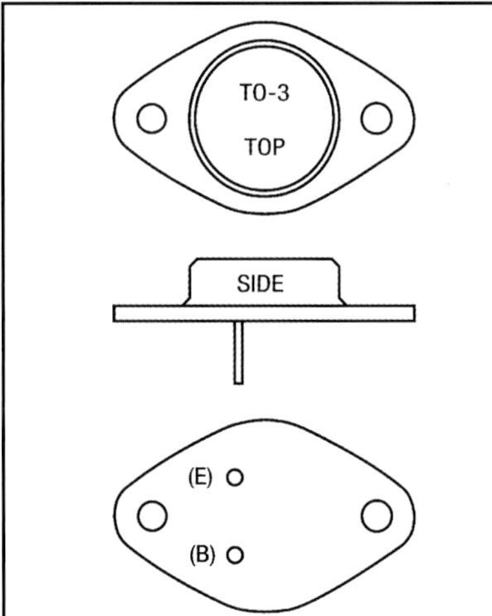

Figure 4-23*: Top, side, and bottom of a typical 2N3055 transistor. (E) marks the emitter lead, (B) the base. The metal case of the transistor is the collector terminal. Use care when mounting the transistor so as not to cause an unintentional short circuit.*

Transistors are manufactured in a variety of physical packages. Among the most common is the so-called TO-3 package. The TO-3 is a metal can with two mounting ears. Two metal legs protrude from the bottom. One of the legs is the emitter terminal, and the other leg is the base. The case itself is the collector.

Figure 4-23 identifies the terminals of a 2N3055 transistor. If you elect to use a transistor other than the 2N3055, it may appear in a different physical form. Even if it's a TO-3 package, the terminal layout may not be the same. Always refer to that part's data sheet to properly characterize it and to identify its terminals.

Part of the reason for all of the exposed metal in these types of transistors is that, in some applications, they're are expected to carry so much current that they heat up. Heat is destructive to transistors, so it must be shed and dissipated somehow. This is accomplished by mounting the transistors to large blocks of metal called heat sinks. The large metal surfaces on power transistors allow heat to flow and be dispersed into the heat sinks. Properly constructed, the Christmas Motor does not burden its transistors enough to cause them to overheat, so no heat sink was needed.

Figure 4-24 depicts the actual circuit diagram for the Christmas motor. The left half of the circuit provides power to right half. The right half, the circuitry contained inside the dotted box, is meant to be replicated six times, once for each reed switch/solenoid pair in the motor. Each instance is connected to appropriate wires on the left-hand circuit. For example, the "BIAS" wire in each instance is connected to the BIAS wire of the circuit on the left, the "GND" wire in each instance is connected to the "GND" wire to the left, and so forth.

In essence, figure 4-24 echoes the circuitry in Figure 4-22, but with crucial additions. Those additions bear explanation.

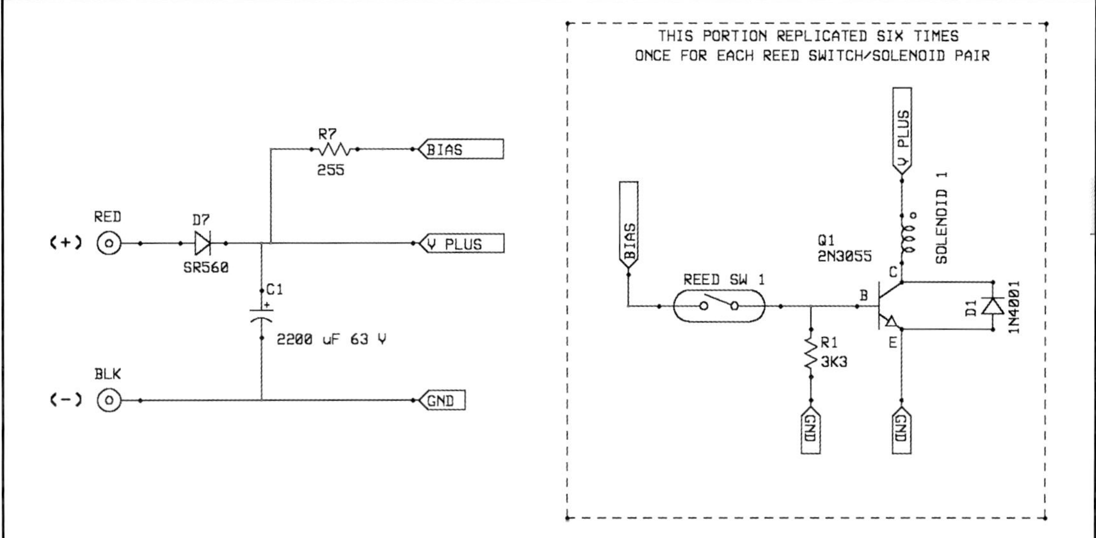

Figure 4-24: *The schematic for the Christmas motor. The circuitry in the dotted box is replicated six times, once for each reed switch/solenoid pair. Only one instance is shown here in the interest of clarity. The circuitry on the left is shared by all six of the other right-hand circuits.*

First, notice that while the base of each transistor (remember—there are six of them) is fed through its own reed switch, ultimately, all six reed switches are sourced through a common 255Ω resistor, designated R7. The purpose of this resistor is to limit how hard the transistors are driven into the "on" condition when a given reed switch is closed. Drive the base of a transistor too hard, and it's possible to damage it.

The 255Ω value is somewhat non-standard, and as the value is non-critical, anything in that ballpark is likely to work just fine. An excellent way to achieve 255Ω is to connect two 510Ω resistors in parallel. Not only is 510Ω a common value, but the use of two resistors in this fashion allows them to share the burden of dissipating the heat produced while they are in operation. With a 24V supply on the motor, I estimate this to be on the order of 2W.

Notice also that the base of each power transistor is connected to ground through a 3.3 kilohm (3,300Ω) resistor (this is designated R1). The purpose of this part is to make sure the base of the transistor is pulled to the same potential as the emitter whenever the corresponding reed switch is open. This assures that the transistor is truly off when it is supposed to be, and not left floating in some partially-on state.

As demonstrated in the earlier motors in this book, the collapse of the magnetic field in a solenoid when the power is cut off can generate high potentials across its terminals. The voltage produced can be high enough to exceed the electrical ratings of a power transistor, and even destroy it. Because the polarity of this high voltage is always opposite the voltage used to energize the coil, this trait can be exploited to suppress it.

Each solenoid in the Christmas Motor is shunted with a diode, part number 1N4001. This is designated D1. A diode is a component that conducts electricity in one direction, but blocks the flow of current in the other. It is oriented in such a manner that when the coil is energized, the diode does not conduct and for all practical purposes, may as well not be present. When the coil is de-energized, however, the collapse of the field produces a high voltage pulse of opposite polarity. In that case, the diode turns on. In doing so, it short circuits the pulse and the energy it contained is dissipated as heat.

The coil driver circuitry includes a bit of protective circuitry, which is comprised of another diode (D7) and an electrolytic capacitor (C1).

Recall that when I described the operation of transistors, I remarked that applying voltages "backward," that is to say, of an incorrect polarity, could result in damage. To prevent this, the positive power line to the coil driver is fitted with a diode, D7, which prevents the application of power to the driver circuitry if the polarity is incorrect. Engineers sometimes refer to this safety feature as an "idiot diode". The diode I used is an unmarked unit salvaged from a defunct PC power supply. Diodes like the 1N5400 or similar components would work just fine.

The firing of multiple solenoids results in a pattern of current consumption that is non-continuous and pulse-like in nature. If the motor is powered by batteries, this is of little consequence. On the other hand, I have observed that some electronic, regulated power supplies have serious problems running the Christmas Motor. Capacitor C1, a 2200 uF, 63V part, is included to "smooth" out the load that the Christmas Motor presents to its power source. Since its inclusion in the driver circuitry, I've observed no problems running the Christmas Motor from any of the power supplies I have at my disposal.

The value of C1 is non-critical, and anything in the ballpark is likely to work. Capacitors of this type can be purchased or salvaged from discarded electronic equipment.

There are two details, with regard to this capacitor, that warrant attention. First, this capacitor is polarized. That means that one terminal is a "plus" terminal, and the other is a "minus" terminal. Care must be exercised to ensure that such a part is wired in the circuitry properly. Connecting this capacitor in reverse can result in damage to the part, a short circuit, or even a loud bang as the case ruptures and the contents of the capacitor fly out. Not good.

Capacitors carry a voltage rating. This voltage should be at least twice the largest voltage that you intend to apply to the motor. Since the design maximum for the Christmas motor is roughly 24V, any substitution made for capacitor C1 should be rated at 50V or greater.

Testing and Results

The Christmas Motor is an interesting and entertaining machine to watch. I have demonstrated to numerous persons, all of whom have delighted in its aeronautical theme, the hypnotic swing of its crank as observed through the viewing port, and the sewing-machine-like chatter of its parts in motion. I have found it to be well worth the effort in constructing it.

Static photographs don't do this machine justice. Even to the naked eye, much of this machine's charm is concealed within the crankcase. Figures 4-25 and 4-26 are models of the motor with crankcase and front bearing support rendered invisible. These are among my favorite illustrations in this book.

The motor never fails to self-start, if sufficient voltage is applied. The motor begins to turn over, grudgingly perhaps, at around 10 or 11V. I have run the motor at as high as 30V, where the propeller turns at over 800 RPM. The motor does a respectable amount of work, as evidenced by the breeze produced by the propeller.

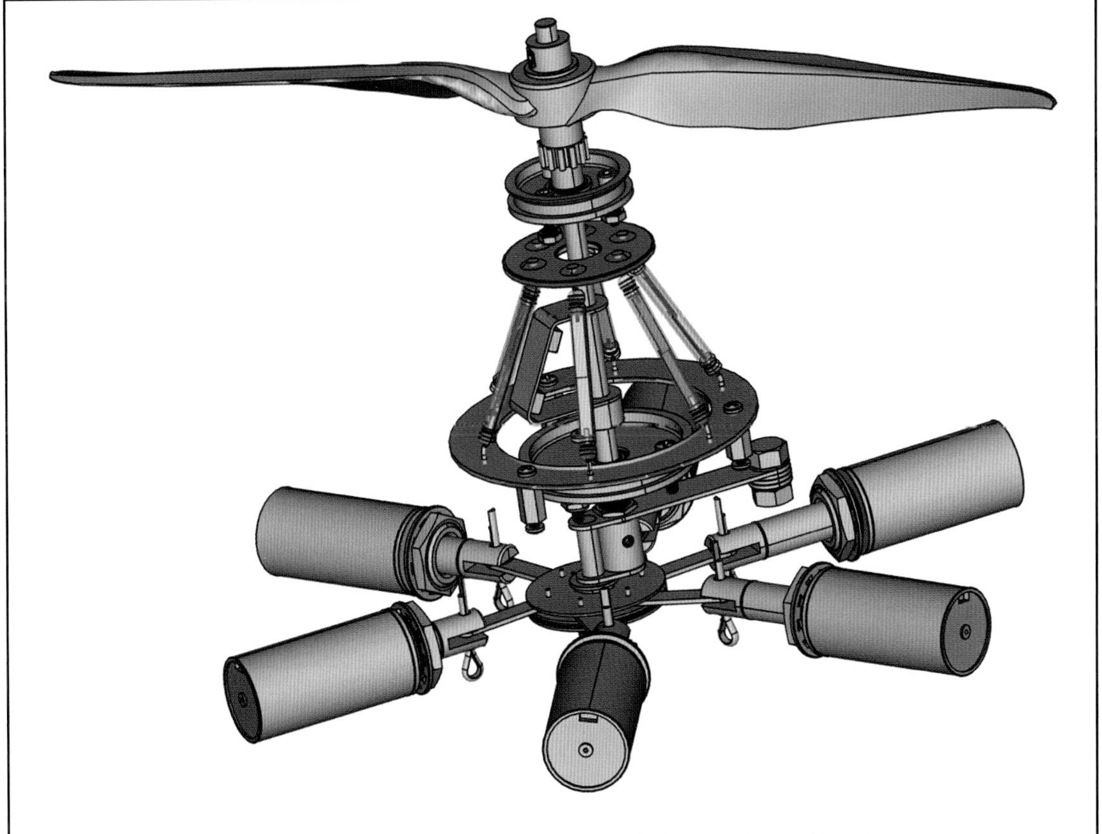

***Figure 4-25**: A "glamor" shot of the motor's internals, with the crankcase, front bearing support, and base structure rendered invisible.*

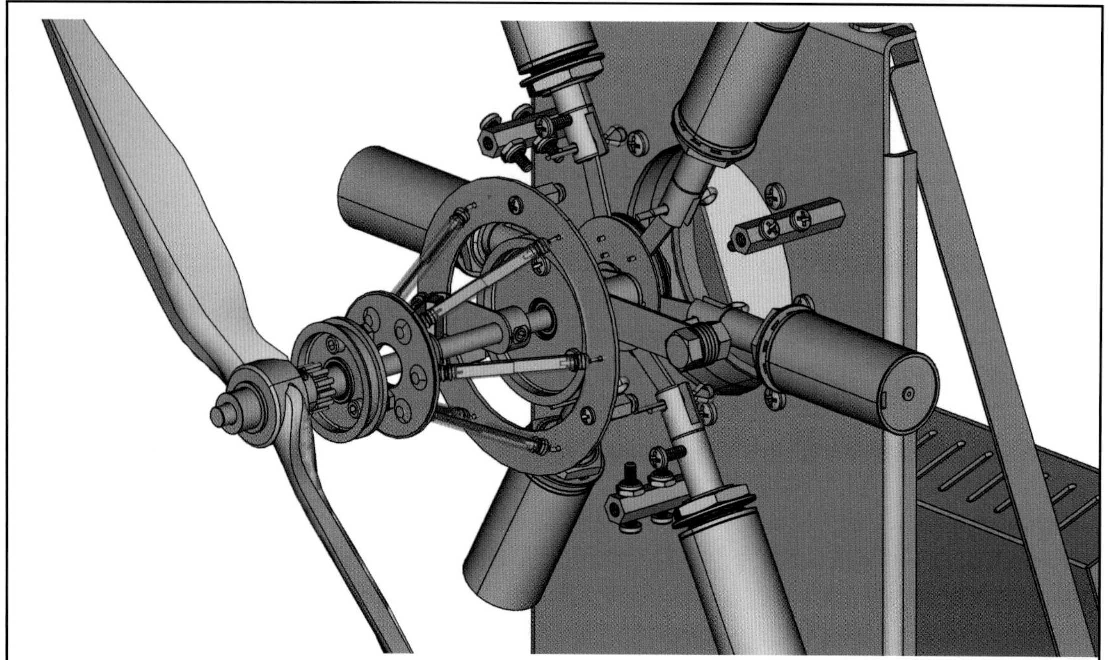

***Figure 4-26**: Another view, sans crankcase and front bearing support, celebrating the intricate internal mechanism.*

In using solenoids in this fashion, I stumbled upon some interesting effects. Initially, the motor ran much more slowly than I would have expected. It was not evident to me why this would be the case, until I carefully considered the design of the solenoids I had used.

In normal use, when a solenoid is energized, the plunger is drawn in, and a force is exerted on whatever mechanism the plunger has been linked to. The resulting motion is acute, jerky, and possibly too violent for some delicate mechanisms. To combat this, to "soften" the draw of the solenoid, designers can make use of the air trapped inside of the coil. As the solenoid plunger is drawn into the bore, it compresses the air inside, like a piston in an engine cylinder. In short order, this pressure is relieved as the air leaks between the plunger and the bore of the coil, but the resistance posed by the air as it escapes acts to slow down the movement of the plunger.

Designers can add little pinholes or vents at the ends of the coil bore, and size them to regulate the rate at which trapped air can escape. This, in turn, limits the speed at which the plungers can be drawn inward.

I discovered that this was the case with my solenoids. In fact, as the motor operated, I became aware of little puffs of air escaping from these holes. Reasoning that the restriction of air movement was limiting the speed of my motor, I used a small drill bit to carefully enlarge these holes. The improvement in motor speed was dramatic. I suspect that higher operating speeds might be obtained if I enlarged the holes further.

Another interesting characteristic of the motor becomes apparent upon inspection of the speed chart in figure 4-27. In the range of 10 to 20V, speed rises in a more or less uniform fashion. Then, as higher voltages are applied, the motor suddenly accelerates.

I have no definite explanation for this behavior, though I have noticed that the "bend" in the curve can move up and down a bit, depending upon whether one reaches the bend through a series of increasing voltages, or decreasing voltages. The solenoid plungers are lubricated with a film of light machine oil, and it's my suspicion that the speed effect is connected to changes in the oil's viscosity as the internal parts begin to warm up.

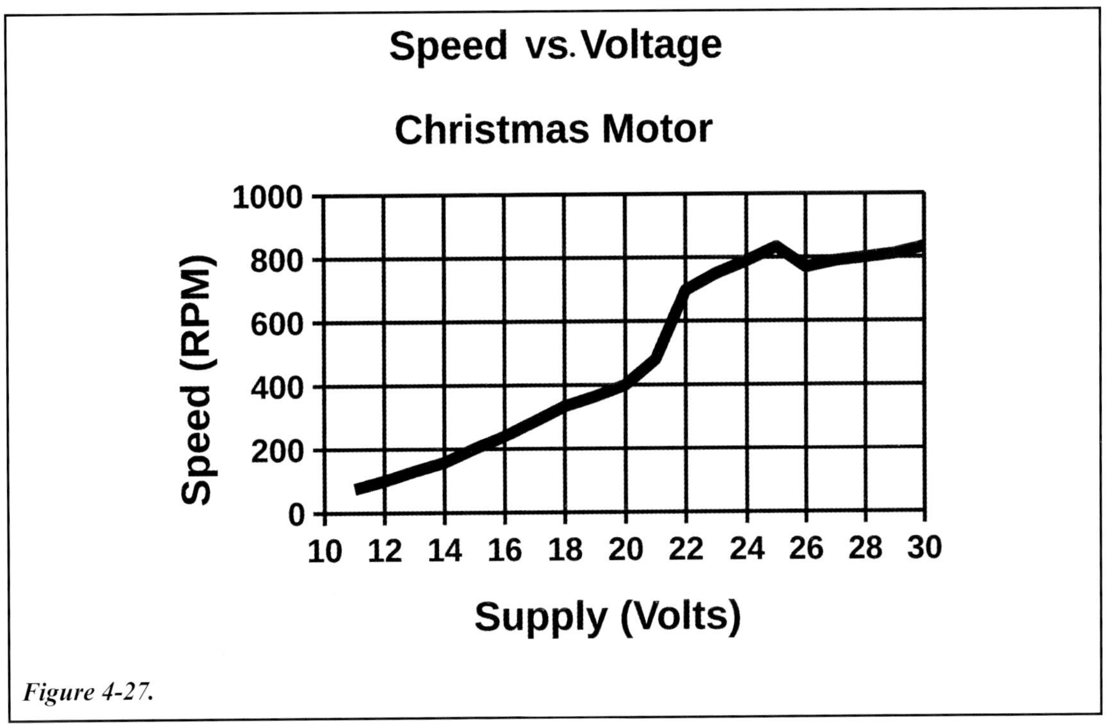

Figure 4-27.

Critical Considerations

Overall, I'm pleased with the Christmas Motor. That said, I can envision a minor improvement or two, and some avenue for further experimentation.

In the coil driver circuit depicted in figure 4-24, all of the reed switches are fed through a single, shared, 255Ω resistor. This is permissible because, in the motor's present configuration, only one reed switch is stimulated to close at any given time. Given the benefit of hindsight, one could argue that a better circuit would provide for each reed switch to be supplied through its own 255Ω resistor. Adding the additional resistors effectively isolates each channel of the coil driver circuit from the others. The revised schematic appears in figure 4-28. The question might be, "*Why bother?*"

Recall that the reed switches in this motor are triggered by magnets mounted on a rotating commutator arm. By tinkering with the size, quantity, and placement of magnets, it is possible to widen the magnets' field of influence and thereby increase the period of time over which a given solenoid is energized. With the revised circuit implemented, one is then free to increase the "on" period to the point that the power stroke of adjacent solenoids will overlap. Having done this, one could expect more robust self-starting behavior, greater running torque, and smoother motor operation.

The swinging counterweight in the crankcase of the Christmas Motor helped to reduce vibration a great deal, though it failed to nullify it entirely. It would be useful to tear back into the motor at some point and add additional mass to the counterweight. This counterweight can be clearly seen at the 3-o'clock position in figure 4-26.

Finally, I can think of one potentially desirable aesthetic improvement: The interior of the crankcase is somewhat dark, despite the window. It would be nice to mount a small light bulb or LED inside the crankcase to light its interior when the motor is in operation. This would make the moving parts inside much easier to see.

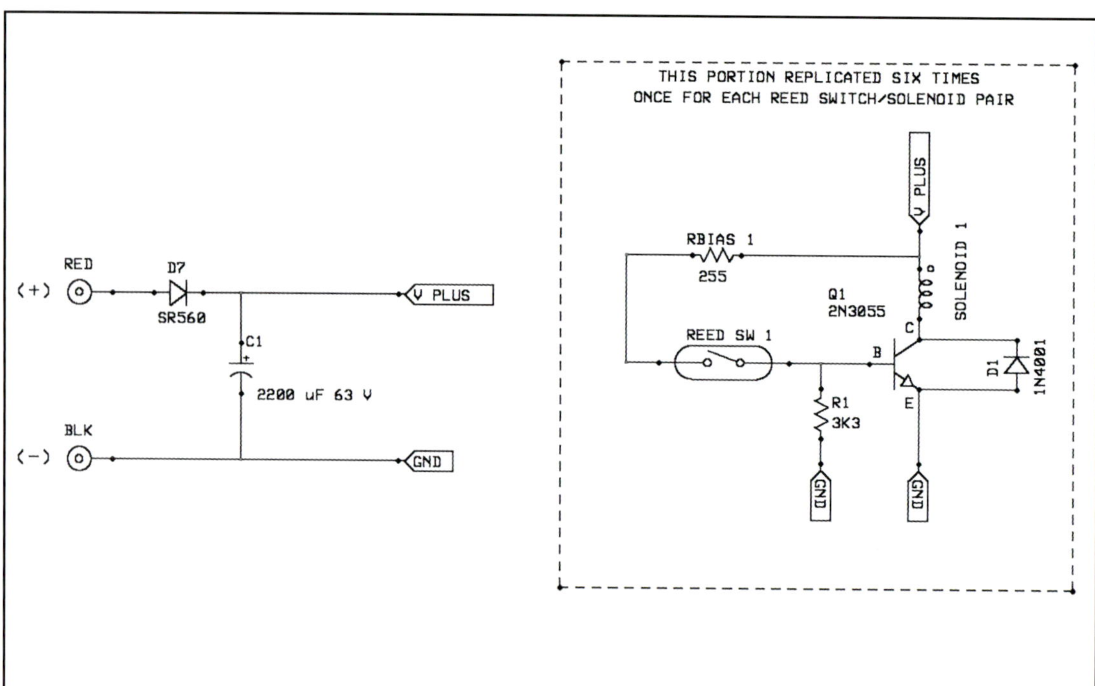

Figure 4-28: *An improved circuit for the Christmas motor. R7, which previously provided a common bias source for all the solenoid drivers, is eliminated. Instead, each driver receives its own bias resistor (RBIAS).*

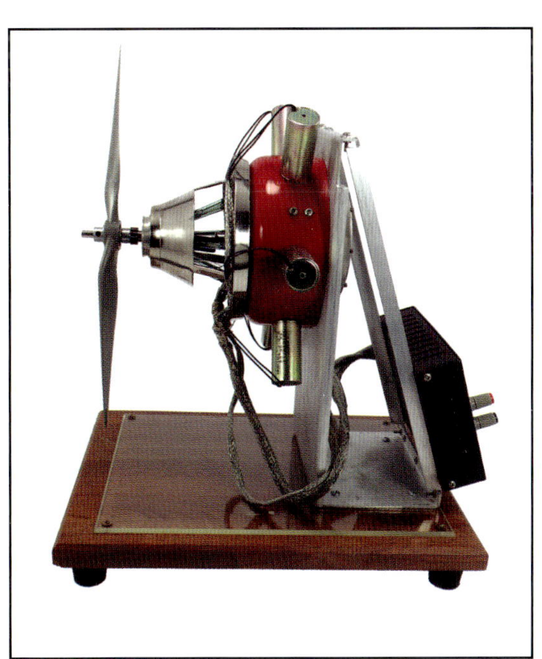

Chapter V
The "Little Twister" Motor

***Figure 5-1**: The "Little Twister" Motor, front.*

Solenoid motors are interesting to build and a lot of fun to watch when they run, but there are good reasons why they're never seen used in any practical application. As we've already observed, they tend not to be self-starting unless extraordinary design steps are taken. Reciprocating plungers and connecting rods sling a lot of mass around, which invariably results in vibration. Perhaps most importantly, these motors tend to be anemic where torque is concerned. This, of course, is an undesired byproduct of crank geometry and other factors that have already been discussed.

The motor described in this chapter, one I like to call the "Little Twister," differs from my prior motors in two important respects.

First, the forces produced by a set of electromagnets are applied directly to a spinning rotor. There is no crankshaft, no connecting links, and no plungers, which means the motor runs quietly, quickly, and with little vibration. In mechanical terms, it's set up like a turbine, as opposed to a piston engine.

Second, the Little Twister makes different use of its electromagnets. The prior motors in this book have relied on the attraction between an energized coil and a piece of steel. The Little Twister, on the other hand, takes advantage of magnetic repulsion between an energized coil and a permanent magnet.

Set under it's glass bell jar, the Twister bears the quaint look of some kind of Victorian laboratory equipment. As you'll discover shortly, however, the principles behind its operation are the same as those that turn the cooling fans in a modern desktop computer. See figures 5-1 and 5-2.

Figure 5-2: *The "Little Twister" Motor, rear.*

The Motor Base

The base for the Little Twister began life as a small table lamp. The lamp was discarded many years ago, long before I set out to build motors, but it's base appealed to me for some reason. So, before relegating the lamp to the trash, I dismantled it and kept the base in anticipation of a project like this one.

The lamp base is a hollow cast metal cube, measuring roughly 4-inches on each side, and 4-inches tall. The foot of the base is somewhat wider, and is embellished with mitered corners and ornamental trim. The exterior has an antique brass finish to it. It's heavy and stable, good attributes for a base of a motor.

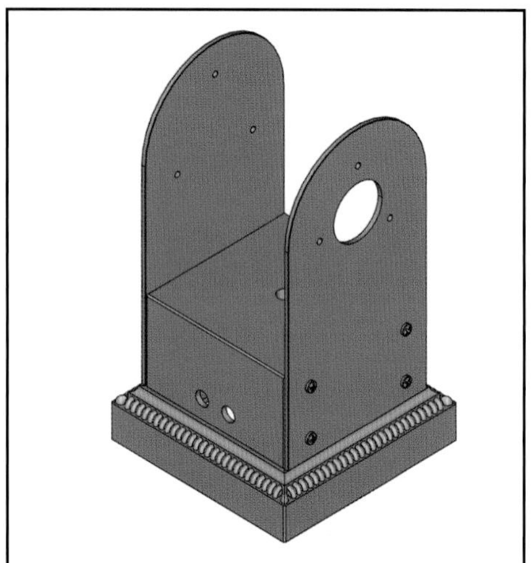

***Figure 5-3**: End plates fashioned from scrap aluminum, bolted to the base of a discarded table lamp, create a functional and attractive frame for the Twister motor.*

Starting with a chunk of 0.125-inch (⅛-inch) aluminum plate, I cut two rectangular end-plates, each measuring 4-inches wide by 7-inches tall. Using 6-32 bolts, washers, and nuts, I bolted these to opposite ends of the base, so as to produce a U-shaped frame. This can be seen in figure 5-3.

In consideration of aesthetics, I decided to remove the end-plates and rework them a little bit to improve their appearance.

I used a compass to scribe a semicircle at one end of each plate. Then, using the scratch mark as a guide, I cut and filed away material so as to produce a pleasing, rounded, "cathedral" window shape.

Since the material from which the end plates were fabricated was scrap material, it was covered with dings, scratches, and other marks. I considered painting the end-plates to conceal these defects, but decided instead to employ a decorative effect called "engine turning" or "jeweling". Possibly the most famous example of engine turning is the treatment given to the engine cowling of Charles Lindbergh's aircraft, the *Spirit of St. Louis*.

To jewel the end plates, I first created a jeweling bit. I started with a ¼-inch flat-headed bolt. Using scissors, I cut a small square of material from a synthetic scouring pad, wrapped it around the head of the bolt, and secured it in place with a plastic zip-tie (a short length of stiff wire will work as well). Next, I inserted the bolt into the chuck of my drill press and gave it a spin. When a bit like this is brought into contact with an aluminum plate, it scratches a circular swirl into the metal. Smaller swirls can be produced with bits made from wooden dowel, or even abrasive "ink" erasers.

Jeweling is a repetitive process, and involves laying down a sequence of overlapping swirls. Swirls are laid down in a single row. When a row is complete, the next row is started, and positioned such that the new row's swirls partially overlap the swirls of the prior row. It's not rocket science, but the effect is most striking if care is taken to assure uniform spacing between individual swirls and between rows.

The effect can be seen in figure 5-4, though the photo really doesn't do it justice. The treatment conceals all but the worst defects in the metal, and in bright light, gives the machinery a sparkling, gem-like appearance.

The Rotor and Bearings

The rotor, the only moving part in the Twister, is the portion of this machine responsible for generating torque on the motor's output shaft. It began life as the wax-capturing dish at the bottom of a set of votive candle holders. The dish is brass, measures about 4-½-inches in diameter, and is about ¾-inch deep. I found a matching pair of them for a dollar at a thrift store.

The outer edges of the rotor dish terminates in a flat lip around its circumference. I drilled three equally-spaced holes around this lip, and used these as mounting points for three flat, ceramic, permanent magnets. The magnets I used measure 1-inch by ¾-inch, and are perhaps 0.187-inch (3⁄16-inch) thick. These magnets were purchased as surplus for a few dollars, but identical ones can be harvested from magnetic catches, of the type used to secure the doors on kitchen cabinets.

In the center of each magnet is a hole. Using 6-32 bolts, nuts, and washers, I was able to attach the magnets to the rim of the rotor. Care must be exercised in tightening the nuts and bolts, because excessive pressure can cause the magnets to fracture. See figure 5-5.

***Figure 5-4**: "Engine Turning" is a technique for improving the aesthetics of machinery. If care is taken during the process, metal surfaces are left with a beautiful, sparkly, gem-like appearance.*

The large, flat faces of these magnets are its magnetic poles. One face is the north magnetic pole, the other, its south pole. The design of the Twister requires that the exposed face of the mounted magnets be south in polarity. One can deduce/confirm this by observing the interaction between a common magnetic compass and the rotor magnets. Recall that where magnets are concerned, unalike poles attract. If the north end of a compass needle is attracted to the face of a rotor magnet, the rotor magnet by definition must be presenting its south pole face. If the compass needle is instead repelled, that particular rotor magnet should be flipped over.

Since the rotor, by definition, rotates, this brings us again to the matter of bearings. So far we've discussed sources for bushings and sleeve bearings, and potential sources for ball bearings. The Twister makes use of some exceptional ball bearings salvaged from an unlikely source—a discarded VHS video tape player. Let me detour for just a moment to explain...

While the ability to record video on magnetic tape has existed since the late 1950's, it was the invention of the VHS video cassette that triggered an explosion in the sale of consumer-grade video cassette recording machines. VHS remained a dominant entertainment delivery format for nearly two decades. Despite having been rendered obsolete by DVDs, Internet streaming, and other digital options, VHS tape recorders are still manufactured and sold. Obsolescence notwithstanding, the fact that a tape machine can store a two hour movie on a cassette cartridge the size of a paperback book is still an impressive feat, involving a bit of engineering magic that you have to understand to fully appreciate.

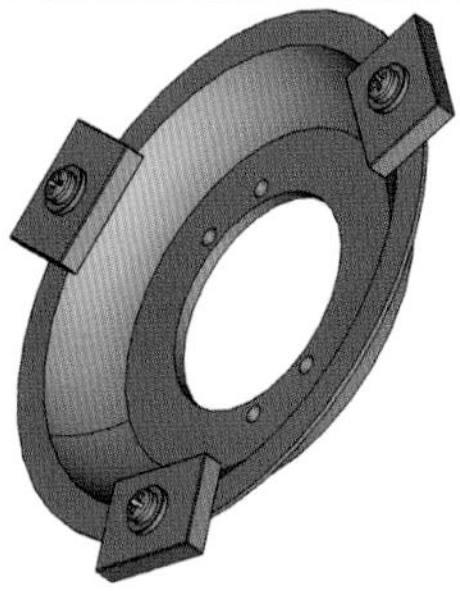

Figure 5-5: The Twister's rotor is comprised of a small brass dish and three magnets of the type used to keep kitchen cabinets closed. The magnets are bolted to the rim of the dish with their south poles exposed.

In recording signals on magnetic tape, there is a relationship between the rate at which data can be stored on the tape, and the rate at which the tape must be dragged across the recording heads. The more information that must be impressed upon the tape in a given span of time, the faster that tape has to move. Since television signals contain huge amounts of rapidly-changing information, the need for absurdly high tape speeds implies the consumption of thousands upon thousands of feet of tape to record even a short television program.

Home video recording was made practical by some out-of-the-box thinking. Instead of simply moving the recording tape against the head, engineers worked out a way to simultaneously move the head against the tape. This trick was accomplished by fabricating the video player's heads in the form of a cylinder or drum which rotates as the tape is drawn past. The effective tape speed of this kind of system is many times the physical tape speed. All VHS tape players and recorders contain such a mechanism, and this was the source of the bearings, output shaft, and mounting hardware used in the Little Twister.

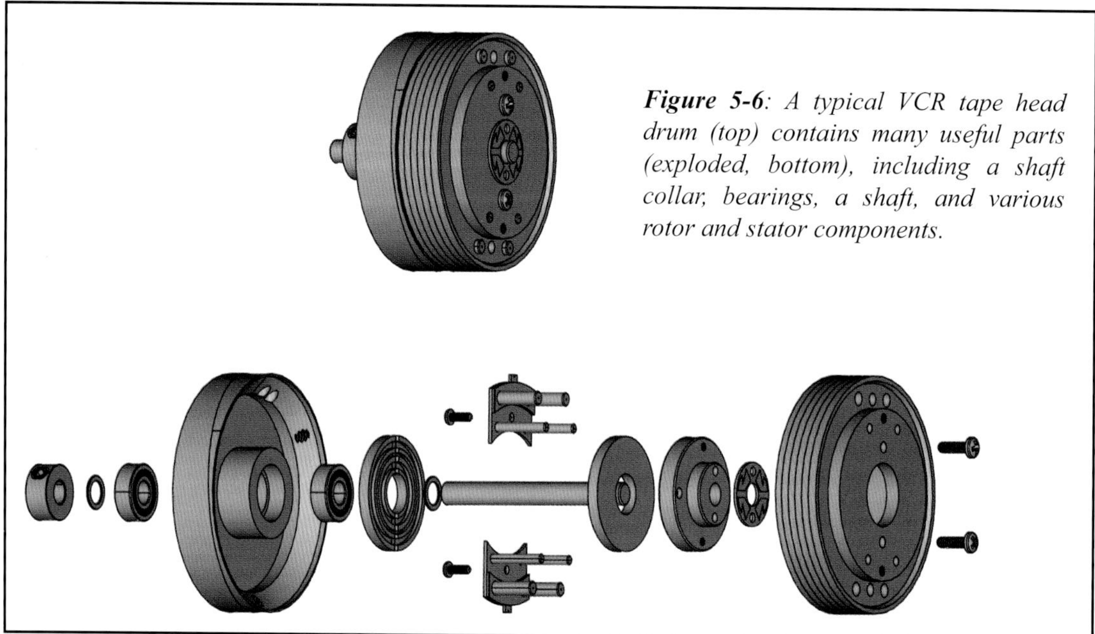

***Figure 5-6**: A typical VCR tape head drum (top) contains many useful parts (exploded, bottom), including a shaft collar, bearings, a shaft, and various rotor and stator components.*

Figure 5-6 shows an example of a drum harvested from a junked video tape player. Note that the drum is actually split around its waist. The right half of the drum, the rotor, contains the actual tape heads. It's the part that spins. It's coupled to a metal shaft that passes through the left half of the drum, the stator. The drum's stator contains a pair of ball bearings that support the shaft and allow it to spin freely. To integrate the drum into the Twister, I took the following steps.

First, I mounted the stator-end of the drum to one of the end-plates attached to the base. I bored a hole through the end-plate, sized large enough for the drum shaft to pass through. Head drums usually sport all kinds of threaded holes which are used to secure them into place in the tape machine. These same holes can be reused (and the original bolts recycled) to mount the drum to the end plate. In my case, I drilled and tapped some fresh holes in the drum, and drilled corresponding holes in the end plate. Three 6-32 screws and lock washers mount the drum to the end plate securely. See figure 5-7.

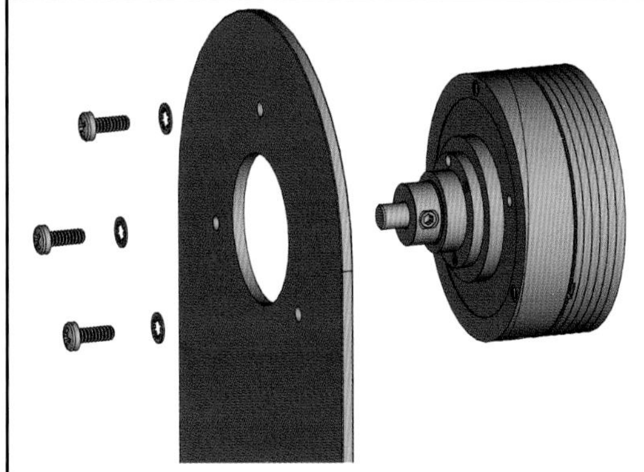

Figure 5-7*: A VCR tape head drum is secured to one of the Twister's end plates. This assembly becomes the bearing on which the Twister's rotor dish will spin.*

The brass rotor dish, described earlier, was fitted to the rotating portion of the tape head drum, and secured with 6-32 screws. As simple as this might sound, there can be challenges. While the tops of some head drums present a flat mounting surface, others have a stepped profile. The drum I used was like this. To get the rotor dish to rest on top, I had to cut a large hole in the floor of the dish. A cylindrical projection at the top of my drum engages this hole, and allows the dish to rest flat against the top of the drum.

I found it easiest to start the dish's retaining screws from the inside of the tape head drum, from which they extend like studs. The dish is secured with a set of four 6-32 nuts and lock washers. Figure 5-8 explains this better than words. In any event, it is essential that the rotor dish be centered on the drum, otherwise the rotor is apt to wobble.

Figure 5-8*: Four 6-32 screws, inserted from the inside of the tape head drum, become studs on which the rotor dish is attached. Four lock washers and nuts keep everything secure.*

The Stator

The rotor, the part of this motor that actually rotates, interacts with and is driven by the stationary part of the motor, called a stator. The stator in this motor is very similar in appearance to the rotor, and is in fact based upon the same kind of brass candle dish that the rotor was made from (recall that in my hunt for suitable materials, described a little earlier, I was fortunate to find a matched set of these dishes). The primary difference between the stator and the rotor is that instead of permanent magnets, the stator is outfitted with coils; six to be exact.

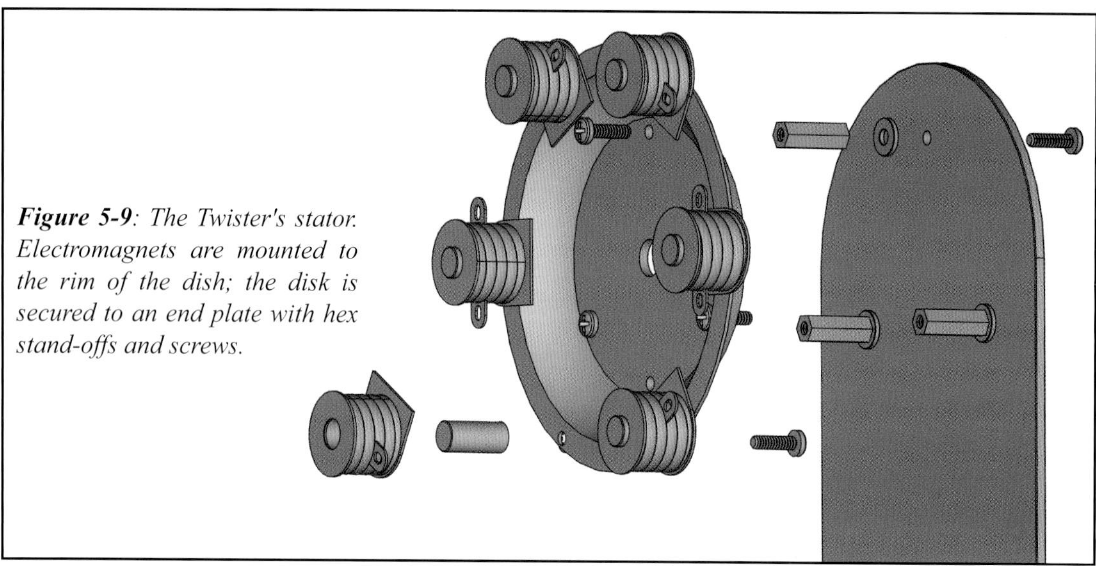

***Figure 5-9**: The Twister's stator. Electromagnets are mounted to the rim of the dish; the disk is secured to an end plate with hex stand-offs and screws.*

I could have fabricated and hand-wound a set of electromagnets as I did on behalf of the Texas Motor, but I decided instead to make use of some recycled junk. I was given a set of burned-out 24V relays. The contacts in these relays were shot, rendering them useless for their intended purpose. I dismantled them and extracted from each a 24V coil and its iron core. The core of each coil was already drilled and threaded, making the electromagnets easy to mount with a single 6-32 screw.

Using a compass, I laid out six holes, equally spaced around the flat lip of the stator dish. I drilled the holes, and then mounted the six coils.

I drilled three more holes in the floor of the dish, equally spaced around the floor, and attached three threaded standoffs. The standoffs I used were about ¾-inch in length. I drilled three corresponding holes in the end-plate mounted to the base, and then used screws to fasten the stator to the end plate. See figure 5-9.

The relationship between the rotor and stator is thus: the rotor and stator face each other, like a pair of opened cymbals. The spacing between the rotor and stator is adjusted so that when the rotor turns, the permanent magnets of the rotor sweep over the pole faces of the coils mounted on the stator. While the rotor obviously must be free to turn without interference, the spacing between rotor and stator is adjusted so that the magnets pass over the coils as closely as possible, without collision. See figure 5-10.

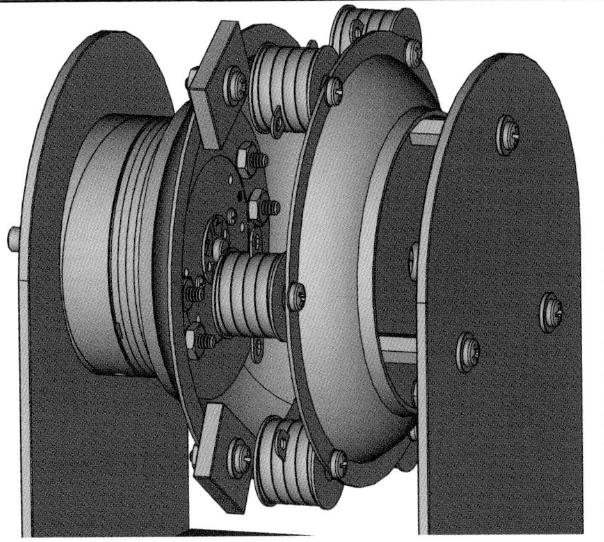

***Figure 5-10**: The spacing between the rotor dish and stator dish is adjusted to minimize the gap between the face of the magnets on the rotor and the core of the electromagnets on the stator. The gap shown here is exaggerated for the sake of clarity.*

The coils are wired together as two groups of three coils each. If we look into the face of the stator, and regard the position of the coils as points on a clock face, the coils at twelve, four, and eight o'clock are wired together. Similarly, the coils at two, six, and ten o'clock form the second group. This idea is depicted in figure 5-11.

While this wiring is very simple, it must be done carefully, and with consideration to the magnetic polarity of the coils. For this design, we'd like all of the coil poles to generate a south magnetic field when energized. To make sure this will be the case, we must identify the terminals of each coil.

A simple way to do this is to connect one of the coils to a flashlight battery or other low-voltage source of direct current. If we bring a magnetic compass in proximity to the pole of the electromagnet, the north end of the compass needle will either be attracted or repelled. The trick is to swap the battery terminals back and forth until attraction is observed. If the north end of the compass needle is attracted to the coil, this implies that the coil's pole face is its south magnetic pole (remember—unalike poles attract). When this condition is achieved, the coil terminal that's connected to the positive terminal of the battery should be marked with a dot of permanent ink or dab of paint. The process is repeated for each of the six coils.

If you turn your attention to figure 5-11 once more, you'll note that the coil groups are wired such that dotted coil terminals are always joined together.

Rotor Position Sensing with the Hall Effect

One could summarize the operation of the Twister Motor this way: by applying electric currents to one or both of the coil groups in the stator, magnetic fields will be produced that will tend to repel the permanent magnets in the rotor. The only "escape" for the permanent magnets is to move laterally, which results in rotation of the rotor and output shaft. However, successful operation is once again predicated on proper timing. The coils need to be energized only when repulsive forces will benefit the rotation of the rotor. At other times, the coils must be turned off. How might this timing be accomplished?

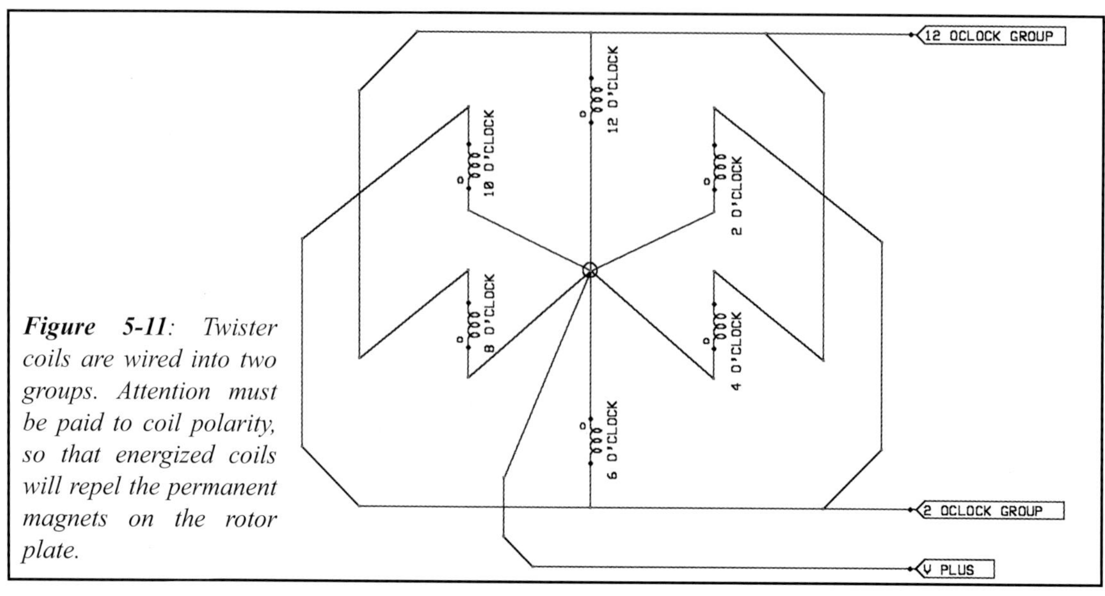

Figure 5-11: Twister coils are wired into two groups. Attention must be paid to coil polarity, so that energized coils will repel the permanent magnets on the rotor plate.

So far, we've seen a couple of examples of cam-driven switches used to control motor timing. That approach could certainly be made to work here. We've also explored one example of the use of magnetic reed switches to drive transistors, which in turn can activate coils. That technique could also have been used here, though I chose to experiment with something different. The timing mechanism of the Twister Motor senses the rotor's position using a device called a Hall effect sensor. The Hall effect and the sensors that capitalize on this property are of sufficient interest that they warrant a brief explanatory detour.

In 1879, Edwin Hall conducted a series of experiments involving a thin gold foil, electric currents, and magnets. One of his discoveries, now known as the Hall effect, is summarized in figure 5-12.

A thin conductive strip is connected to a power source, such that current flows from point (a) to point (b). If this conductive sheet is subjected to the influence of a magnetic field (perpendicular to the surface of the sheet) then a voltage will appear between points (c) and (d). This emergent voltage, called a Hall Voltage, is proportionate to the strength of the applied magnetic field.

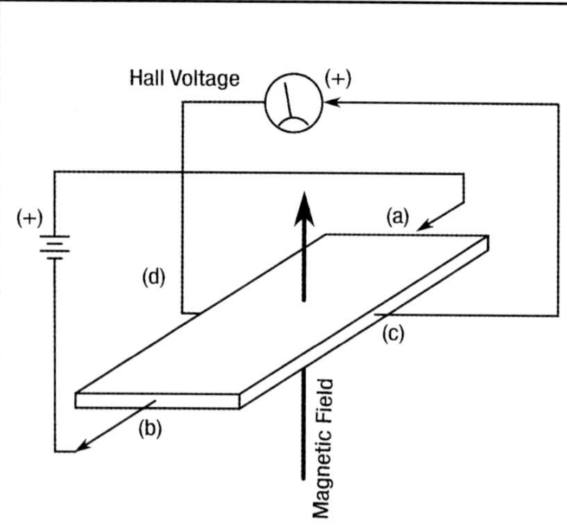

Figure 5-12: When a thin current-carrying conductor is subjected to a perpendicular magnetic field, a small voltage appears at right angles to both. This voltage, called the Hall voltage, is the essence of the Hall effect.

While the Hall effect is a subtle one, modern semiconductor manufacturers have developed practical solid-state sensors that exploit it. These sensors contain built-in amplifiers and signal conditioning circuitry to convert feeble Hall voltages into large, robust signals capable of driving other electronics.

Hall effect sensors come in several flavors. One type is the linear-output sensor. In this class of sensor, the output of the device is a voltage that varies proportionately with the strength of the applied magnetic field. While this type of sensor is useful for all sorts of applications, it's not what we're looking for, here.

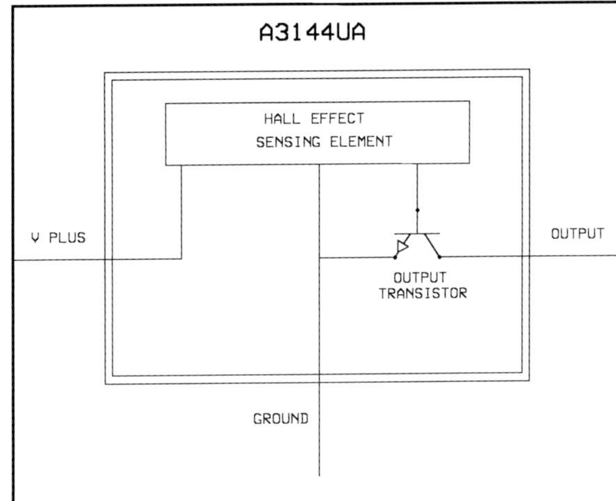

***Figure 5-13**: The A3144UA Hall effect sensor contains a transistor that "pulls" the output pin to ground when the sense element is in proximity to a magnet.*

Another type of Hall device is the digital output sensor, also known as a Hall switch. The digital output sensor contains an internal voltage standard, which is electronically compared to the Hall voltage produced when the sensor encounters a magnetic field. If the field is greater than the threshold represented by the internal standard, the device turns an output transistor on. If the field is less than the threshold represented by the standard, the output transistor is turned off. In essence, this type of Hall sensor behaves like a magnetic reed switch, with the exception that there are no moving parts to wear out. It's just the thing we need to sense the position of Twister's rotor.

The generic Hall switch just described responds to only one pole of an external magnet, either north or south. If the opposite pole is applied to the sensor, the switch does not turn on, regardless of how intense the field may be. For this reason, a Hall switch of this type is regarded as a unipolar switch.

Bipolar Hall switches exist, and their behavior differs somewhat from the device just described. Like the unipolar switch, the output transistor of the bipolar switch is off until the applied magnetic field exceeds some threshold. Then, the output transistor turns on. However, in order to turn the transistor back off, it is not sufficient merely to reduce or eliminate the applied magnetic field. In the bipolar Hall switch, the field must actually be reversed, and that reversed field must also exceed a threshold value.

It is useful, for purposes of analogy, to regard the unipolar Hall effect switch like a doorbell button, that is to say, as long as the button is pushed, the bell rings. When the button is released, the bell stops. A bipolar Hall effect switch, on the other hand, is like a lamp with a pull-chain. Action must be taken to turn the lamp on, and further action must be taken to turn it back off.

The Twister Motor employs a specific model of unipolar Hall effect device, the Allegro A3144UA. The A3144UA activates when in the presence of a reasonably-strong south magnetic field. As the magnets on the rotor plate (which were oriented to present their south magnetic poles) sweep past the A3144UA device, it responds by switching on when a rotor magnet is in close proximity, and off again when the magnet has swept past. In this way, it is possible to "sense" the rotation and relative position of the rotor. That electrical signal can be used to coordinate the firing of the stator coils.

Two A3144UAs are used in the Twister motor. This part was purchased through an online surplus dealer for about fifty cents each. Brand-new units, and numerous comparable parts, can be purchased for around a dollar. I'll discuss some possible alternatives to the A3144UA in just a moment.

The Coil Driver

Each of the A3144UA Hall effect switches used in the Twister contains a tiny output transistor. When the A3144UA is off, that is to say, there's is no magnetic field present to stimulate the sensor, the output transistor is off. In this state, the output pin of the device is regarded as "floating," meaning that, internally, it isn't really connected to anything. When the sensor is stimulated, the output transistor turns on. Internally, the output pin of the device is connected to the ground terminal. See figure 5-13.

Conceptually, we should be able to wire our coils by connecting one terminal to the positive power supply lead, and the other to the output pin of the Hall effect switch. The idea here is that, as a field impinges on the sensor, its internal transistor activates, which would have the effect of connecting the switched end of the coil to ground. The coil should then energize.

This is all well and good, but like the magnetic reed switches discussed in the last chapter, the Hall effect switch has limited drive capability. That is to say, it is not up to the task of powering coils directly. So while the Hall effect device can be used to sense rotor position and effect motor timing, some kind of interface has to be inserted between it and the coils themselves. The coil driver circuity used in the Twister Motor is, in some ways, very much like the driver circuitry that appears in the Christmas Motor. Still, there are some differences that need to be explained. Please refer to figure 5-14.

First, let's discuss the power supply portion of the circuit. Diode D1 and capacitor C1 should look familiar. Diode D1 protects the motor from electrical damage should the power source be connected backwards. Capacitor C1 provides energy storage that helps to suppress some of the noise created by coils that are rapidly switched on and off. Together these two parts result in a supply line that identified as "V PLUS" on the schematic. If you look elsewhere on the schematic, you'll notice that V PLUS is connected to all of the coils. Ultimately, this is the voltage that powers the coils and runs the motor. Note that since the input of the motor can be varied over a wide range of voltages, V PLUS will also vary. Changing the input voltage to the Twister Motor is one way to control the speed of the motor.

The little box marked "U3" is an LM7805 three-terminal voltage regulator. This part takes V-PLUS as its input, and converts whatever voltage V PLUS is to a steady 5V. This is denoted as "5 V PLUS" on the schematic. This regulated source of 5V is used to power the Hall effect switches. Note that the LM7805 can only work its magic over a reasonable set of input values. If "V PLUS" is less than 8V or so, there is not enough overhead for the regulator to function properly, and its output will start to drop off. If V PLUS is above 35V, the regulator can be damaged. Trying to run the Twister at less than 8V is not likely to cause any harm, even if the motor fails to operate, but any motor using a driver based on the Twister circuit should definitely not be run above 35V.

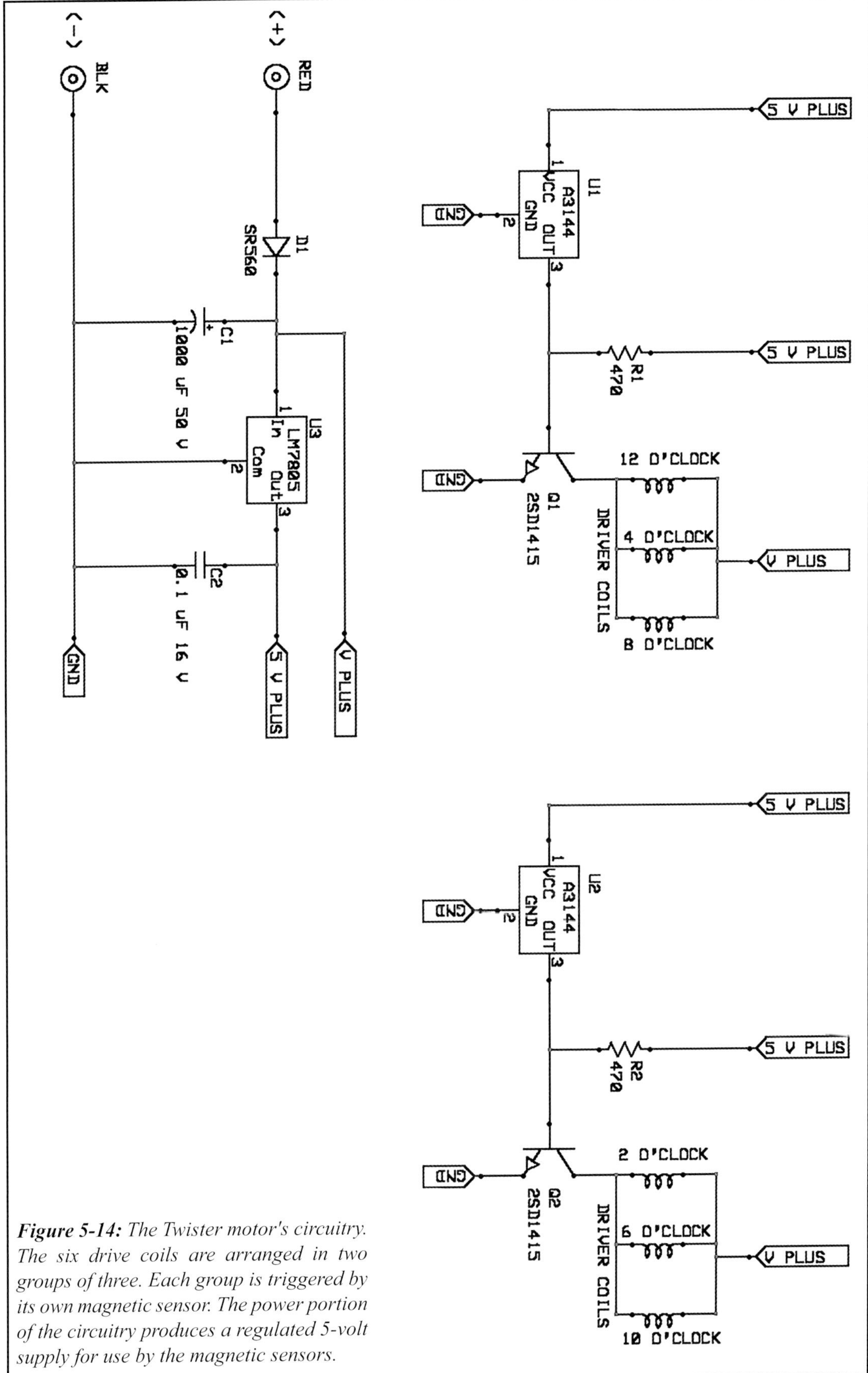

Figure 5-14: *The Twister motor's circuitry. The six drive coils are arranged in two groups of three. Each group is triggered by its own magnetic sensor. The power portion of the circuitry produces a regulated 5-volt supply for use by the magnetic sensors.*

This brings us to the portion of the circuitry that actually drives the coils. There are two identical circuits, one which controls the coils at the twelve, four, and eight o'clock positions, and one which powers the remaining coils. Since these two sections are identical but for the coils they are wired to, I'll focus my explanation on one of them, only.

Transistor Q1 is a 2SD1415 switching transistor. This transistor is a common component in PC (computer) power supplies. The transistors used in the Twister were both harvested from discarded computer hardware.

As far as the coils are concerned, transistor Q1 is wired the same way that the transistors in the Christmas motors were wired. When Q1 turns on, it connects the lower lead of the coils to ground, which energizes the coils.

On the input side of Q1, however, things look a little different. Resistor R1, a 470Ω resistor, ties the base of the transistor to "5 V PLUS." Doing so automatically causes the transistor to turn on, which energizes the coils.

The base of Q1 is also connected to output pin of the Hall effect switch. Recall that when the sensor is unstimulated (no magnet present), the output transistor inside the sensor is off, which leaves the output pin essentially unconnected. Thus, it has no effect on resistor R1's ability to drive transistor Q1, and therefore the coils remain on.

Conversely, when a magnet moves into proximity to the sensor, the Hall effect sensor's internal output transistor pulls the output pin to ground. Because it's also connected to Q1's base terminal, its base is also pulled to ground. This negates the effects of R1, causing transistor Q1 to turn off. As a result, the coils also turn off.

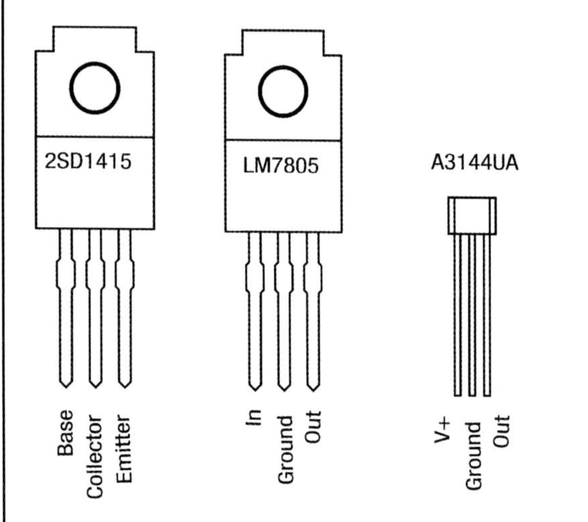

Figure 5-15: *Pin-outs for some of the semiconductors used in the Twister. Substitutions are fine, but when in doubt, always refer to the technical datasheets for the specific parts you intend to use. These sheets are readily available on the Internet.*

Notice that in the Christmas Motor, the natural state of the driver (and therefore, the coils) was "off." Applying a magnet to a reed switch activated the corresponding transistor, which energized the attached coil. In the Twister Motor, things act in reverse: the normal state of the driver is "on". When the Hall effect sensor detects a magnetic field, it deactivates the corresponding transistor to the off-state, which deenergizes the attached coils.

For your reference, figure 5-15 depicts the 2SD1415, the LM7805, and the A3144UA packages and identifies the functions of the parts' various terminals.

Motor Timing

The Hall effect sensors are tiny plastic devices with no obvious means for mounting. Figure 5-16 shows that these sensors can be epoxied to the head of a steel 6-32 flat head screw. The screw not only supplies a means by which a sensor can be supported and installed in the Twister, it provides the added benefit of improving the sensitivity of the sensor. This is due to the tendency of nearby magnetic fields to be drawn through the sensor and into the shaft of the screw.

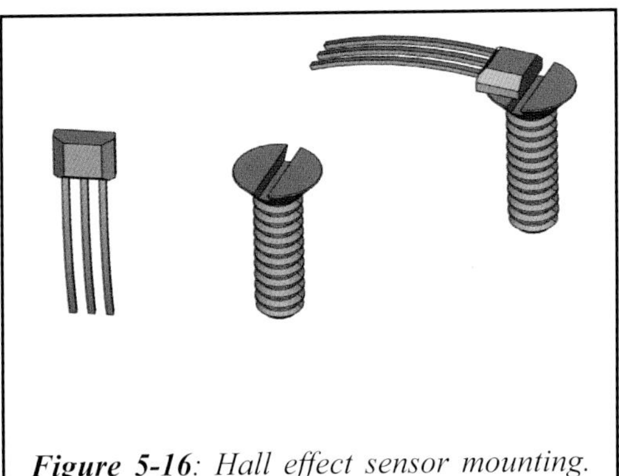

Figure 5-16: *Hall effect sensor mounting. The sensor (left) is glued to the top of a flat-head screw (center). The finished article appears to the right. The Twister Motor requires two of these.*

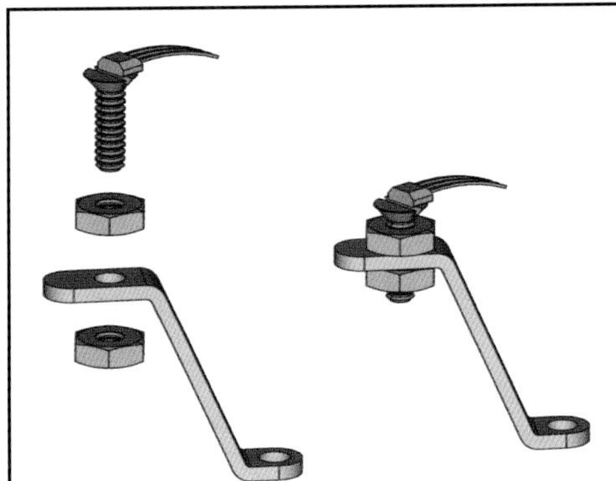

Figure 5-17: *A Z-shaped bracket and a pair of jam nuts supports the Hall effect sensor. On the left, the constituent parts. On the right, a complete sensor/bracket assembly.*

The sensors are held in their proper position by a Z-shaped bracket fashioned from a section of ¼-inch brass strip. A hole was drilled in each end of the "Z." One end was fitted with the sensor-and-screw assembly, and the other end was anchored to the floor of the stator. Jam nuts allow for adjustment of the height of the sensor off the bracket. See figure 5-17.

Figure 5-18 shows the approximate location of the Hall effect switches on the stator dish. Sensor U1, the sensor which controls the coils at twelve, four, and eight o'clock is located near the seven o'clock position. Sensor U2, which controls the other three coils, is located near the five o'clock position.

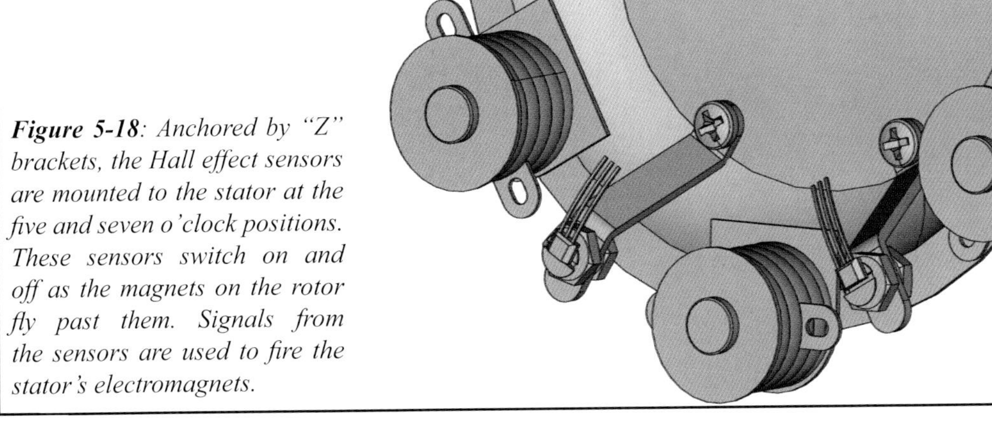

Figure 5-18: *Anchored by "Z" brackets, the Hall effect sensors are mounted to the stator at the five and seven o'clock positions. These sensors switch on and off as the magnets on the rotor fly past them. Signals from the sensors are used to fire the stator's electromagnets.*

Proper motor timing can best be understood in terms of a single sensor, a single coil on the stator, and a single magnet on the rotor. Since the default state of the driver circuitry is to turn the coil on, in general, a magnet on the rotor will be repelled and driven away from a nearby coil. This force results in torque on the rotor, which causes the rotor to turn.

When a magnet passes over a Hall effect switch, this disables the coil controlled by that switch. If the Hall effect switch is positioned such that the rotor magnet must pass over it just before passing over the coil, the coil will be momentarily deactivated, and its repulsive force will be neutralized. The momentum in the rotor will carry the magnet across the electromagnet's pole face.

At some point, the rotor will have turned far enough that the rotor magnet no longer influences the Hall sensor, but is still in proximity to the coil. At this moment, the coil is reenergized, and repulsion forces resume. The rotor is given another kick, and around it goes. See figure 5-18.

Obviously, the Twister Motor is somewhat more sophisticated than this, but the explanation still holds. Instead of one, there are three magnets on the rotor. At any one time, those three magnets can interact with three coils on the stator. For this reason, coils on the stator are wired in groups of three.

In fact, the Twister uses two sensors and two groups of coils, each with its own sensor, to act on the rotor magnets six times for every revolution of the rotor. The result is a motor that is inherently self-starting and one which can generate torque no matter what the angular position of the rotor.

Getting the Twister to run is simple, but it does require some essential tinkering. You must bear three important factors in mind:

Factor #1: The Hall effect switches which turn the coils on and off are triggered by the permanent magnets on the rotor. Since the A3144UA is a unipolar device, it can only be activated if the applied field is of the proper polarity. If you build your own Twister, and find that the Hall effect switches don't seem to be working, double-check your wiring. If that survives scrutiny, then you will need to remove the permanent magnets from the rotor and flip them over. This will present the opposite pole to the Hall effect switch.

Factor #2: This motor relies on magnetic repulsion to generate its motion. That means that when the coils in the stator are on, the magnetic pole produced by each coil must be the same polarity as the pole of the magnets on the rotor. If the energized coils do not repel the permanent magnets on the rotor, the wires feeding the coil or coil set must be reversed.

Factor #3: As always, optimal performance is realized through carefully-adjusted timing. The rotor will always turn in a direction that brings the magnets across the Hall effect switches before their associated coils. The best placement of the sensors depends upon the shape and size of the magnets on the rotor. A well-designed sensor bracket will allow for some adjust-ability.

The Flywheel

The Twister Motor really doesn't require a flywheel, at least not in the sense that attraction motors like the Peewee and Texas motors do. Still, the motor shaft offered by the VCR head amounts to little more than a stub. While the motion of the rotor is easily visible when the Twister is in operation, the motion of a naked shaft is not so obvious. I decided to augment the shaft with a bit of decoration that I euphemistically refer to as a "flywheel."

This object, in fact, is nothing more than the base of an old brass candle holder, inside of which I soldered a shaft coupler. The flywheel is ornate and interesting to watch as it turns. The shaft coupler and its set screw allows me to easily attach (and remove, if necessary) the flywheel to the shaft stub. See figure 5-19.

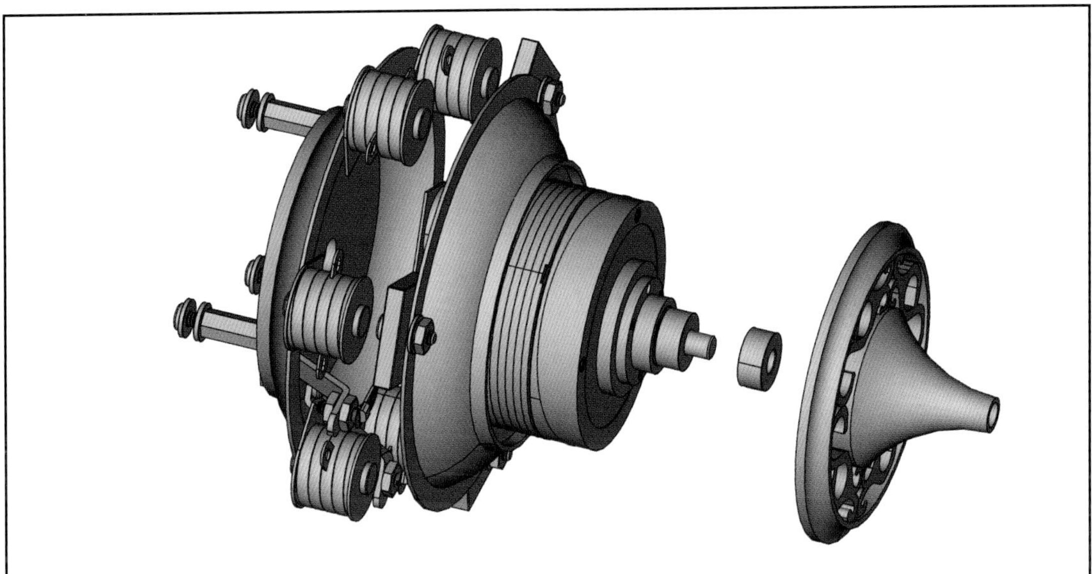

***Figure 5-19**: A shaft coupler, soldered to the interior of a brass candle holder base, creates an aesthetically pleasing flywheel.*

Testing and Results

Performance of the Twister Motor is outstanding. As indicated earlier, it is inherently self-starting. It produces torque all the way through each revolution, so the motor starts quickly, speeds up quickly, and runs very smoothly. With this motor, there is no clatter and none of the sewing-machine-like sounds characteristic of reciprocating attraction motors. The motor is virtually silent in operation, with the exception of the sound the permanent magnets make as they slice through the air.

The VCR head bearings on which the rotor turns are as smooth as butter. If the motor is run up and then disconnected from its power source, it takes a long, long time before the rotor finally comes to rest. You would be hard pressed to find a finer set of bearings for a small machine like this.

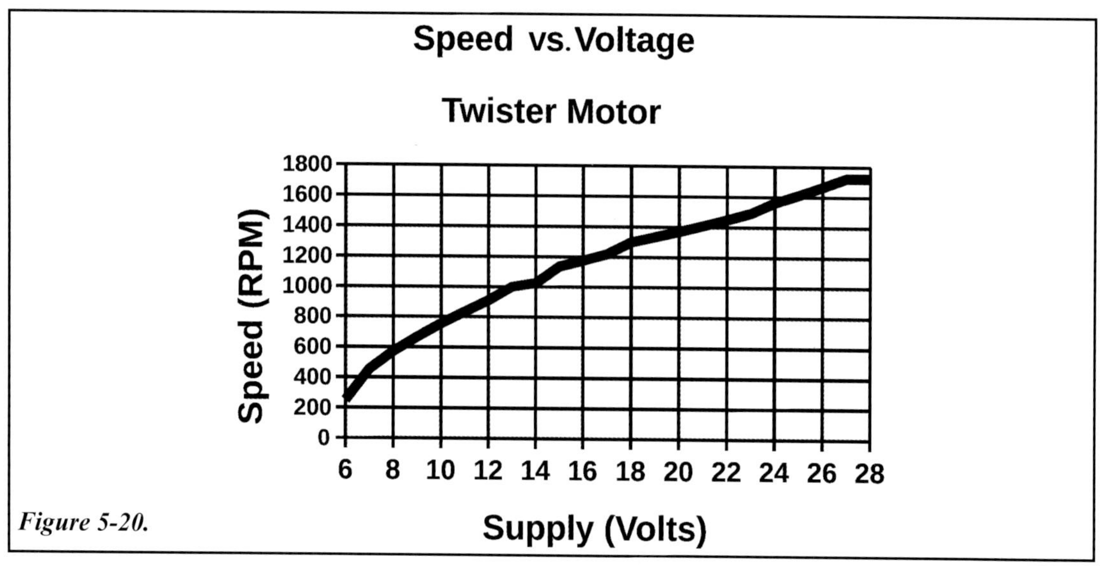

Figure 5-20.

If started manually with a spin from one's fingers, motor operation can be sustained on as little as 5.5V. It becomes self-starting between 6.5 and 7.0V. At 29.0V, the highest voltage tested, the motor spins like a little tornado at a respectable 1,720 RPM. Figure 5-20 shows the relationship between operating voltage and shaft speed.

Critical Considerations

Among the motors presented so far, the Little Twister is most adaptable to a size or geometry that is capable of doing real work.

If there is anything bad to be said about the Little Twister, it's that it exhibits some vibration that really shouldn't be present. It's safe to say that poor materials can yield poor results, and that the brass which makes up the rotor and flywheel of the Twister are the shoddy products of third-world craftsmen. The rotor dish is slightly out-of-round, and the thickness of its cross section varies in a non-uniform manner. The vibration these defects produce occurs despite my best efforts to assure concentricity and balance.

I took several steps to try to improve the balance of the rotor, including the use of a drill bit to bore a series of shallow dimples in the base. The hope was that better balance might be achieved by removing excess brass from "heavy" areas. This technique is rather common where small rotating machine parts need to be balanced.

Ironically, experimentation revealed that by careful adjustment of the flywheel on the shaft, it was possible to partially cancel the imbalance in the rotor with the imbalance inherent in the flywheel. This was done by adjusting the "heavy" part of the flywheel to lie 180 degrees from the "heavy" part of the rotor. In the end, vibration is still present, but it's much less noticeable.

Another potential downside to a motor of this type is the fact that the direction in which the rotor turns cannot be changed on the fly, owing to the fact that rotor direction is determined by the physical placement of the Hall effect switches.

If switch-selectable bidirectional operation is required, one could add another set of Hall effect switches, positioned to enable rotor rotation in the opposite direction, and use a toggle switch to allow the operator to select which set of Hall sensors will trigger the driver circuitry.

Speaking of Hall sensors, the A3144UA part used in this motor was acquired as surplus, and as such, it may not be available when it comes time for you to build your own Little Twister. This should be no problem, because there are many other types of sensors that should work just fine. Any substitute should be a unipolar Hall sensor with an open-collector (digital) output. A quick internet search of brand-new, currently-manufactured, components yields the SS441A (Honeywell), ATS137-PL-B-B (Diodes, Inc), and the A1101LUA-T (Allegro Microsystems). They should all work just fine in this application, and none cost more than a buck or two apiece. If a suitable part can be found in surplus, it's likely to be even cheaper.

Hall sensors can be found in certain kinds of electronic scrap. I developed the circuit for the Twister using a Hall Sensor that I pulled out of an old ultrasonic fish finder. Hall sensors may also be found in certain printers, copiers, fax machines and scanners, in electronic motors that drive VCR machines, in archaic computer peripherals like floppy drives, and even in the solid-state fans that cool computer cabinets. Be advised that some of these salvaged Hall devices are probably bipolar in nature, and won't work in the Little Twister unless changes are made. Recall that bipolar sensors require exposure to both magnetic poles in order to toggle from the "on" condition to the "off." This requirement might be met by changing the rotor to accommodate four permanent magnets instead of three, and arranging them so that their magnetic poles alternate. Of course, such a change to the rotor will require corresponding changes to the number and position of coils on the stator.

My final comment concerns the 2SD1415 transistors used to drive the motor's coils. While this part is depicted in my schematic diagram as a simple NPN transistor, its internals are actually quite a bit more sophisticated. The 2SD1415 is a so-called "Darlington" transistor, the import of this being that the 2SD1415 can be induced to switch on and off with comparatively weak signals applied to its base terminal. For applications of this type, this characteristic is a good thing.

As noted above, the 2SD1415 can be found in discarded computer power supplies. Note that if the 2SD1415 is not present, the power transistors in these supplies will surely have similar characteristics. It is always prudent, however, to try to find a manufacturer's data sheet on the part in question to help identify the terminals and to verify that the electrical characteristics of the part will meet your requirements.

The clever reader will recall that the driver circuitry for the Christmas Motor featured a set of diodes which were used to suppress the high-voltage spikes generated by solenoids when their fields collapse. At first glance, these protective diodes seem to be absent in the Little Twister's circuitry. Actually, they're not. Another feature of the 2SD1415 is that diode protection is built right into the transistor itself! If some other transistor is used in place of the 2SD1415, and the manufacturer's data sheet for the alternate part does not depict an internal protective diode, external diodes should be added.

13-4-36

Chapter VI
The "Copper Queen" Motor

Figure 6-1: *The "Copper Queen" Motor, front.*

Just about everyone has heard of Tombstone, Arizona, the site Wyatt Earp's infamous gun battle at the O.K. Corral. Ironically, it seems that few have ever heard of the town of Bisbee, about 25 miles to the south, though one could argue that it is a place of far greater importance.

Every single one of the devices we associate with modern life — cars, refrigerators, computers, television, radio, electronic medical devices, robots — not to mention our Marvelous Magnetic Machines, contain copper wire, and lots of it. Copper is mined from the earth, so the importance of places where copper-bearing ores are plentiful cannot possibly be overstated.

Bisbee was born at the site of one of the richest copper deposits ever found on Earth, and work in Bisbee mines from the late 1870s to the early 1970s has yielded more than 8 billion pounds of copper. Early on, Bisbee was regarded as the "Queen of the Copper Camps", and among its most productive mines was the celebrated "Copper Queen".

This chapter concerns the final motor described in this book. It is a self-starting rotary motor with a permanent magnet field, and a three-pole rotor. Its construction bears a certain Victorian look that reminds me of the old mine machinery on display in front of Bisbee's Mining and Historical Museum. Because of this, and because of the conspicuous use of copper in the rotor coils of this machine, I refer to it, too, as a "Copper Queen".

Figure 6-2*: The "Copper Queen" Motor, rear.*

The Motor Frame and Base

In searching for a frame for this machine, my initial inclination was to employ an old aluminum pot or saucepan, as I had done in building the Christmas Motor. No doubt, that approach had served me well. Yet one day, while poking through the kitchen items in a second-hand shop, I came upon a decorative fruit basket, composed of thin steel wire segments that had been artfully bent and spot-welded into a bowl-shaped container. I was about to cast it aside when I thought the better of it. It seemed to have sufficient structural integrity to become a motor frame, and it offered one advantage that a simple sauce pan could not — an unobstructed view of the motor's internals.

Having decided on the use of the wire basket as the frame for my machine, my first problem became apparent. I envisioned the Copper Queen to be a horizontal-shaft motor. The basket would be set on its side, and the bearings that support the axle of the motor would be mounted in the floor of the bowl. Yet, the floor of the basket was comprised of little more than a ring of steel wire, and a whole lot of air. There simply was no surface upon which bearings could be mounted. As it turns out, the solution was comparatively simple.

First, I acquired two disks of metal, each about 4-½ inches in diameter. The first disk was cut from scrap aluminum sheet, about 0.050-inch thick. The second disk was a shallow brass saucer, purchased at a yard sale for a quarter. A series of six holes, drilled through both disks in a circular pattern, allows the use of 6-32 screws and nuts to join the two disks together, like a sandwich.

I placed the aluminum disk against the floor of the basket from the inside, and the saucer against the floor of the basket from the outside. I installed six screws, lock washers and nuts, and then tightened them securely. The disks clamp against the wire hoop in the floor of the basket, and form a rigid, flat surface upon which bearings can be installed later. See figure 6-3.

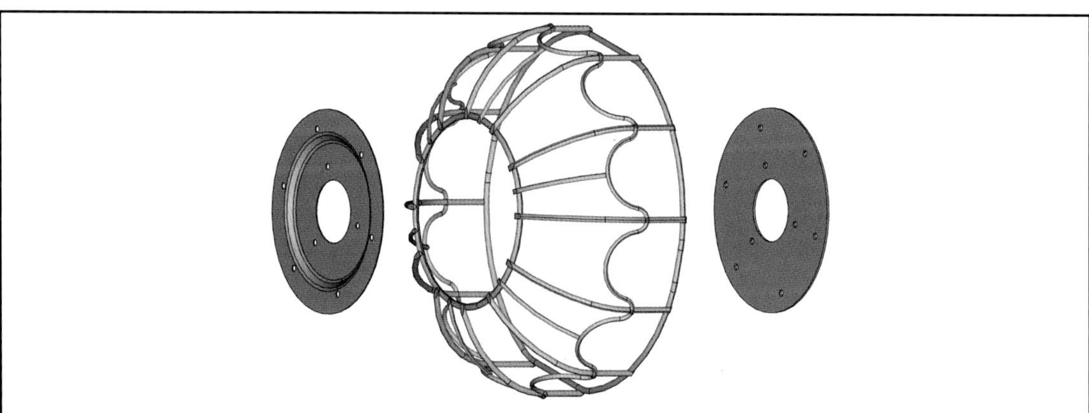

***Figure 6-3**: A shallow brass dish and an aluminum disk create a floor for the Copper Queen's wire-basket frame.*

The basket, by itself, would be of little use without a suitable base or pedestal on which to mount it. This need was answered through the acquisition of an old paper towel dispenser.

The dispenser I found was the counter-top variety, comprised of a thick hardwood dowel mounted to a heavy, circular, cast-aluminum base. In this type of dispenser, the dowel projects vertically from the base, forming a spindle upon which a paper towel roll can be placed.

I purchased the dispenser at a church sale for a buck, took it home, and removed the dowel. I noted that the harvested base had a verdigris finish that somehow complements the finish on the wire basket. It also features a leaf pattern, cast into the aluminum. In terms of visual composition, the leaves offer a pleasing contrast to the contours of the wire in the bowl.

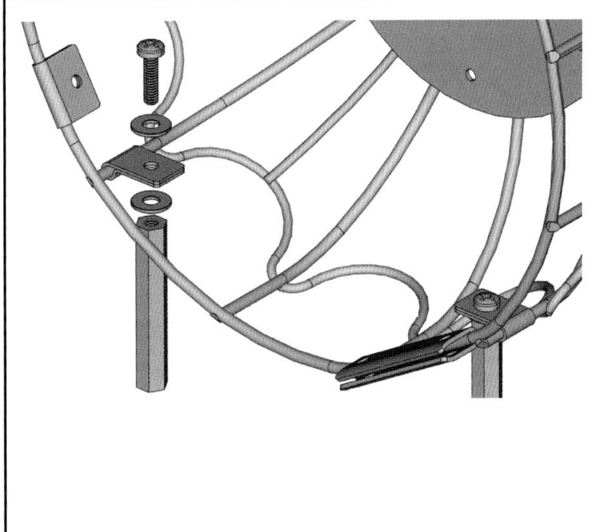

***Figure 6-4**: Small strips of metal, bent into a C-shape and wrapped around the basket wires, create tabs to which legs and other accessories can be mounted.*

As noted in my description of the Christmas Motor, it's easy to mount something to an aluminum sauce pan. It's simply a matter of drilling a hole and attaching what you wish with a small bolt and nut. The skeletal nature of the fruit basket in the Copper Queen, on the other hand, complicates matters, as there is nothing tangible in which to drill a hole.

Through some experimentation, I arrived at a design for little metal tabs that, when fitted to the wire elements of the basket at any desired location, can be clamped into place with the tightening of a screw. These tabs consist of nothing more than small strips of heavy brass or aluminum sheet metal, which are formed in a flattened C-shape. The legs are drilled through to provide passage for a bolt. See figure 6-4.

The fruit-basket frame was mounted to the paper-towel-dispenser base with three hexagonal aluminum standoffs, which act like the legs of a tripod. Two of the legs support the rear of the motor and are attached to the wall of the fruit basket with the C-shaped tabs or clamps I've just described. The third leg supports the front of the motor. It's attached through the use of a little brass L-bracket, to one of the screws that was used to clamp the disks to the floor of the basket. All three legs are then bolted to the base with 8-32 screws. See figure 6-5.

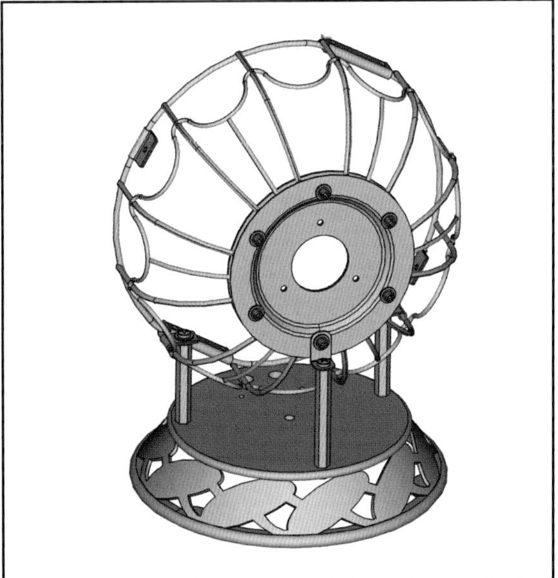

***Figure 6-5**: The Copper Queen's frame is attached to the base with three legs, comprised of hex stand-offs.*

The Motor Bearings

It should be evident, by this point, that bearings are a crucial component of any rotating machinery. Over the course of the last few chapters, I've suggested various alternatives for bearings, and more than one of them would have made serviceable bearings for this project. That said, I was sufficiently impressed with the performance of the VCR head bearings, used in the Little Twister Motor, to employ them again in this project.

DVD and DVR systems have rendered the VHS tape recorder/player all but obsolete, so there is no shortage of equipment from which heads can be extracted. I regularly see fully functional second-hand decks sell for five or ten dollars, and broken units are virtually always free.

After harvesting the head drum unit from a donor tape deck, I removed a pair of screws at the top to split the assembly in two (figure 5-6, in the prior chapter, shows a representative VCR tape head drum assembly in exploded form). With the screws removed, the top, rotating, part of the drum came free. The bottom half, containing the bearings, shaft and other hardware, is the stator or base of the drum assembly. This is the part we're interested in at the moment. See figure 6-6.

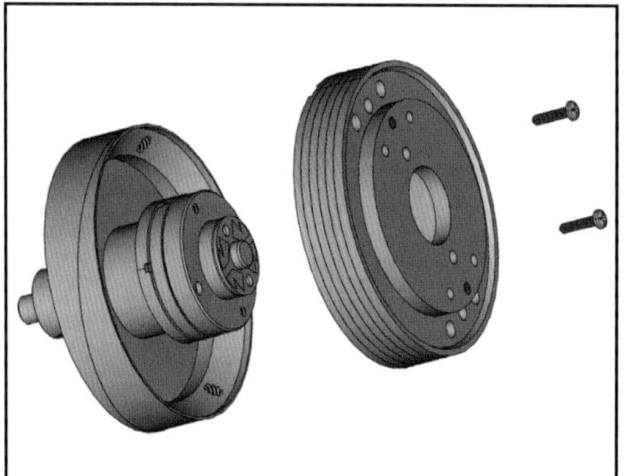

Figure 6-6: *A VCR tape head assembly with the rotor removed (right). On the left is the stator or base of the assembly, which contains bearings, a shaft, and a flanged hub to which the rotor is normally attached.*

The metal disks that were clamped into the floor of the wire basket frame provide a nice flat surface into which the lower (stator) portion of the drum can be mounted. Note that the mounting surface of the drum itself is not flat. The drum is around 2-½-inches in diameter, but steps down to a shallow neck just under 1-½-inches across. To mate with this, I bored a hole through the center of the disks in the floor of the basket, sized to fit the neck on the drum. I inserted the drum from the inside of the basket, and secured it to the disks with three screws. A cross section of the wire basket, disks, and installed bearings undergoing assembly can be seen in figure 6-7.

Since head drums are manufactured in Asia, the threaded holes and screws that fit them are all metric. It's good practice, when stripping any old piece of equipment, to save any screws you remove.

In the case that the original screws are too short, or otherwise unsuitable, there is no reason why the threaded holes in the drum can't be drilled out and re-tapped as needed. Here in the United States, the preponderance of readily-available fasteners still have English threads and sizes. I drilled and tapped the mounting holes in the base of my head drum to accept 6-32 screws. This made sense, because I have a collection of several thousand assorted 6-32 screws in various lengths and with differing head styles.

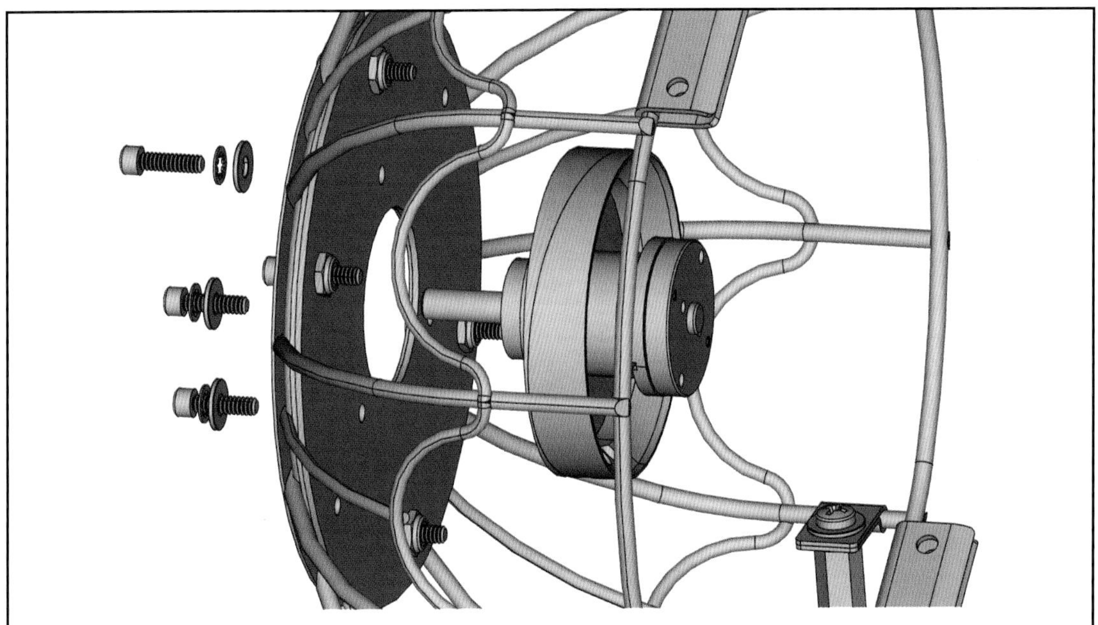

***Figure 6-7**: Here, the lower half of the VCR head assembly — containing bearings and shaft — is being loaded into the Copper Queen's frame. The assembly is secured to the plates in the front of the basket with three 6-32 screws, lock-washers, and washers.*

The Rotor

Construction of the rotor for the Copper Queen begins with the top-half of the tape head drum. Normally its concave side faces the shaft to form its enclosed-drum-shape. The preferred geometry for the Copper Queen rotor is the opposite — where the rotor is flipped and its open end faces away from the shaft.

On most of the drums I've come across, the top half of the drum is attached with screws to a brass disk or crown which has been press-fit onto the shaft. If these screws are removed, the top half of the drum can be detached from the shaft, inverted, and then reinstalled with the same screws. The transition from the stock to inverted form is depicted in figure 6-8.

The rotor of the Copper Queen features three large electromagnets, which are mounted along the circumference of the drum. The coils are spaced equally, 120 degrees apart.

The electromagnets I used consist of solid, soft-iron cores pressed into plastic bobbins measuring 1-inch in diameter, and about 1-¼-inches in length. One end of the core in each coil has been drilled and tapped to accept a number-8 machine screw. This is the means by which the electromagnet can be mounted to a surface.

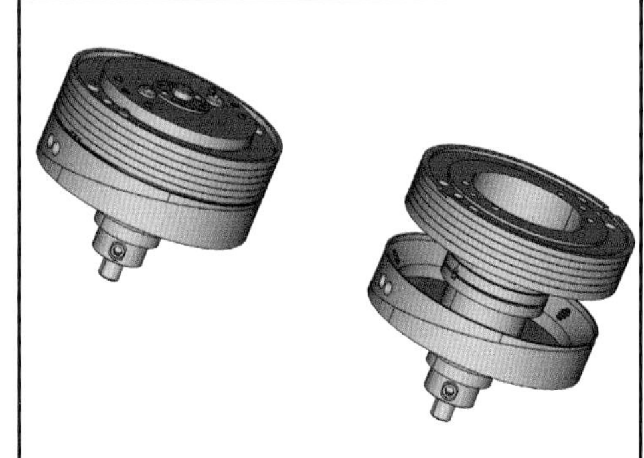

***Figure 6-8**: Two typical VCR head drums. The assembly on the left is stock. On the right, superfluous innards of the drum have been removed and the drum's rotor has been removed, inverted, and reattached.*

I am uncertain as to the original purpose of these coils, though they look like they may have been harvested from some old doorbells. Similar coils can be found in large relays or contactors. One could certainly wind a set of coils from scratch, using the instructions I provided in Chapter 3.

Mounting the electromagnets requires further modification to the rotor drum. Six holes, equally spaced 60 degrees apart, must be drilled along the periphery of the drum. Three of the holes are sized to pass a number-8 screw. The other three holes must be made large enough to clear the heads of the screws used to fasten the electromagnets to the drum. Figure 6-9 shows how each bolt is inserted through a hole opposite the position where the corresponding electromagnet will be mounted.

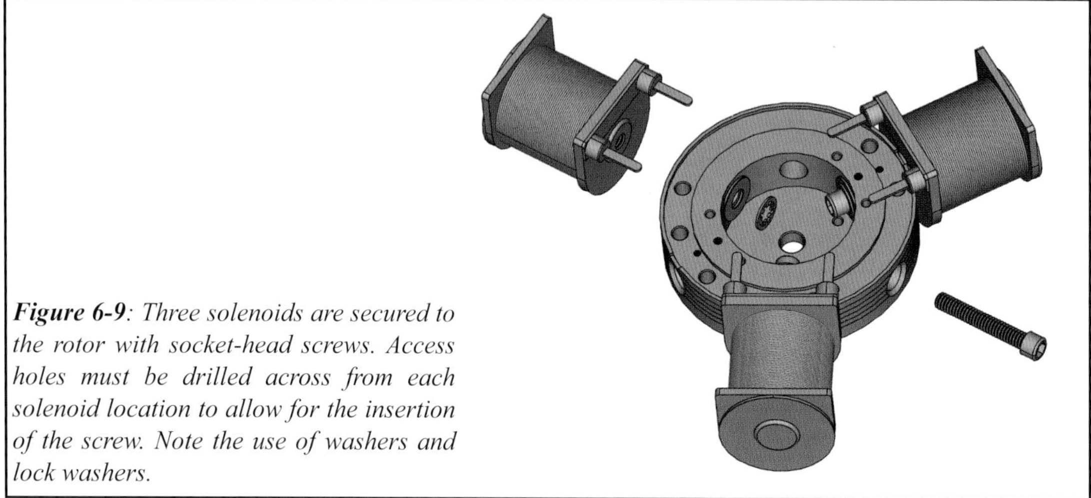

***Figure 6-9**: Three solenoids are secured to the rotor with socket-head screws. Access holes must be drilled across from each solenoid location to allow for the insertion of the screw. Note the use of washers and lock washers.*

Flathead and even Phillips-head screws are poor choices for this kind of job. I used socket-head screws and lock washers to fasten the electromagnets to the rotor.

In Chapter 5, I described a method of using a battery and a magnetic compass to characterize a coil to determine its polarity. The same technique should be executed here. Apply a small voltage to the terminals of one of the rotor coils. Swap the battery terminals, as necessary, so as to cause a magnetic compass needle to point towards the coil's exposed pole face when the coil is energized. When you've achieved this, look to see which coil terminal is positive (connected to the positive terminal of the battery) and mark it with a dot of ink or paint. Repeat until you have characterized all three coils.

The Commutator

Mounting coils on a moving rotor poses a challenge. How does one deliver electricity to the coils? And, as with every other motor in this book, there is an issue of timing. It's not sufficient merely to energize the coils. They have to be energized at the correct time. The solution to the problem is answered with a fixture called a commutator.

Commutators can vary in shape and size, but a typical commutator can be seen in figure 6-10. Commutators are cylindrical in shape, and are composed of a core of insulating material. The surface of the cylinder is covered with metallic strips or segments, which run parallel to the commutator's axle. Each segment is electrically insulated from its neighbors, and is wired in some fashion to the coils in the rotor. Depending upon the size of the motor and the number of coils in the rotor, there may be as few as two segments on the commutator, or there may be dozens.

Electricity is delivered to the commutator with a set of sliding contacts called brushes. The brushes rest on the surface of the commutator, and conduct electricity into the segments resting beneath the brushes. Notice that, as the rotor turns, new segments slide beneath the brushes. Thus, the relationship between the sliding brush and the rotating commutator allows energy to be delivered to the rotor while it's turning. The angular relationship between the commutator segments and the coils to which they are wired determines the motor timing. There are many options for brushes. I'll describe the Copper Queen's brushes in greater detail in just a moment.

Figure 6-10: *A typical commutator as found on a small DC motor.*

It is possible to salvage a commutator from another motor and use it in your own version of the Copper Queen. I would expect suitable commutators to be found in scrap motors taken from electric blenders, hand drills, power saws, sewing machines, and other appliances that make use of so-called "universal" motors.

The trouble is that, in each of the examples cited, the shaft of these motors runs the full length of the motor housing. The commutators in these motors are pressed onto the same shaft that the rotor spins on. In the case of the Copper Queen, the shaft does not pass all the way through the motor. Rather, it only projects from the front of the motor. There is no "rear shaft" to mount a conventional commutator upon, so I endeavored to fabricate something special.

Construction of the Copper Queen's commutator began with an aluminum hex standoff (drilled and tapped for a #8 screw) and a piece of ¾-inch copper pipe. The standoff, which measures 1-½ inches in length, is inserted inside of the copper pipe. The pipe was cut to a length of ¾-inch.

Next, I inserted and wedged three plastic spacers between the standoff and the interior of the pipe. By the term "spacer," what I really mean is bits of plastic rod, rigid plastic tubing, pieces of plastic swizzle stick, drinking straw, or small segments cut from a set of chopsticks. It doesn't matter much what you use, provided that your spacers are non-conductive and are of uniform diameter. Uniformity is important to assure that the axis of the standoff is located precisely on the axis of the copper pipe. In other words, the stand-off must be centered and concentric with the pipe.

I plugged one end of the copper pipe with modeling clay, mixed up a batch of epoxy, and poured the liquid into the space between the pipe and the standoff. I left the epoxy to cure for twenty-four hours. When it had hardened, I removed the clay. The result is an epoxy cylinder with an aluminum center and a copper sheath. See figure 6-11.

The Copper Queen design calls for a commutator with three segments. I measured and laid out saw lines on the surface of the copper pipe with an awl. Then, carefully, using a hacksaw blade with as fine a pitch as I could find, I made three cuts through the copper pipe. The cuts run parallel to the axis, and are only made deep enough to separate the copper into three equally-sized segments. You don't want to hack your way too far into the epoxy core. See figure 6-12.

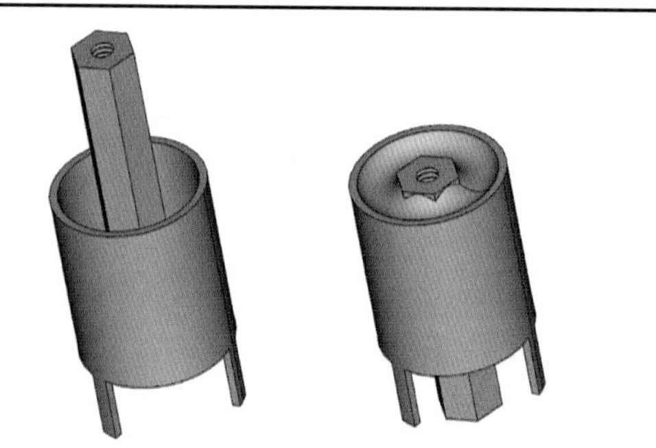

***Figure 6-11**: Details of commutator fabrication. A hex standoff is inserted into a section of copper tubing (left). The standoff is centered in the bore, the bottom of the tube is plugged with modelling clay, and the interior of the tube is filled with epoxy (right). "Legs" cut into one of the tubes will provide tabs to which wires will be soldered later.*

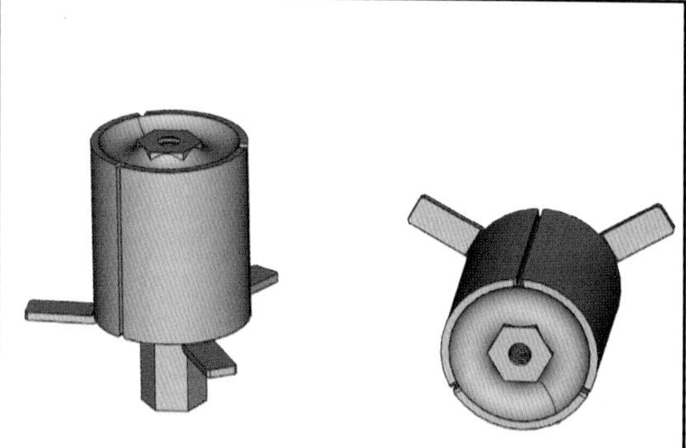

***Figure 6-12**: Two views of a completed commutator. The copper sleeve is separated into three segments with a hacksaw blade. Tabs at the bottom end are bent over to provide better access for soldering wires.*

Later on, wires must be soldered to each segment. This soldering can be done at one edge of the commutator. A better scheme is to start with a copper pipe that is ⅛-inch longer than need be. It can be cut back with a file or nibbling tool to yield three little tabs (visible in figure 6-11) one for each commutator segment. These tabs provide a convenient point to which wires can be soldered. This is a feature I exploited in the Copper Queen, and is depicted in figure 6-12.

The remaining task is contrive some way to attach the commutator to the face of the rotor drum. Actually, this turns out to be simple, and requires no more than an aluminum disk, about 2-½ inches in diameter.

The disk requires seven holes. The first hole is drilled dead center in the middle of the disk. The hole is sized to pass a #8 screw. A screw is passed through this hole and threaded into the standoff in the commutator, to bond the two together. Liberal use of lock washers or thread-lock compound is probably a good idea here.

Three more holes, set near the periphery of the disk and equally spaced, 120 degrees apart, were located to align with holes already present in the top half of the VCR drum.

The precise location of these holes, of course, is dependent upon the specific VCR drum used in the construction of the motor. In the worst case, you may have to drill your own mounting holes in the drum. Using these holes, the disk is bolted to the open face of the drum. I inserted some ¼-inch standoffs to raise the disk off the face of the drum a little bit. See figure 6-13.

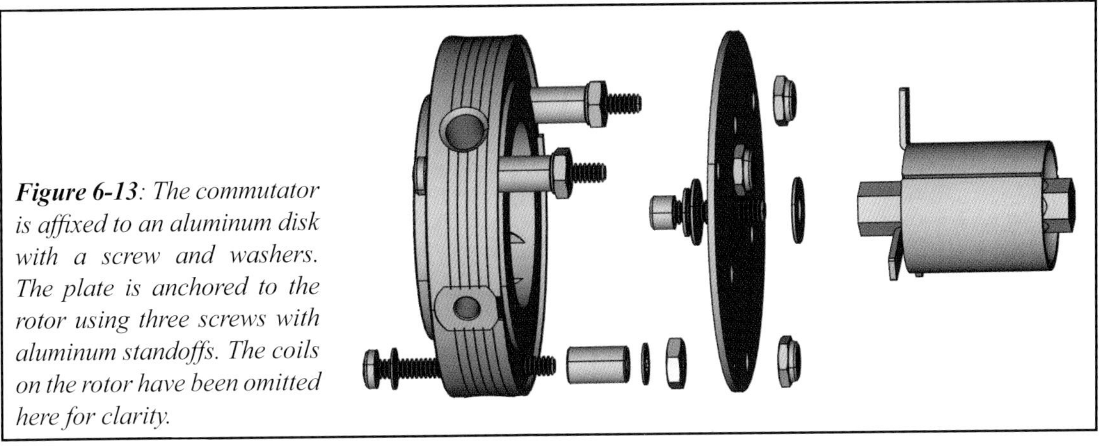

Figure 6-13: *The commutator is affixed to an aluminum disk with a screw and washers. The plate is anchored to the rotor using three screws with aluminum standoffs. The coils on the rotor have been omitted here for clarity.*

The remaining three holes were drilled to allow wires to pass through the disk from the rotor coils to the commutator segments. Since there are three segments to the commutator, there are three wire holes, evenly spaced, 120 degrees apart.

The commutator is wired to the coils as shown in the diagram in figure 6-14. Note that each segment of the commutator is connected to two adjacent coils. When wiring the coils, pay attention to the dotted terminals.

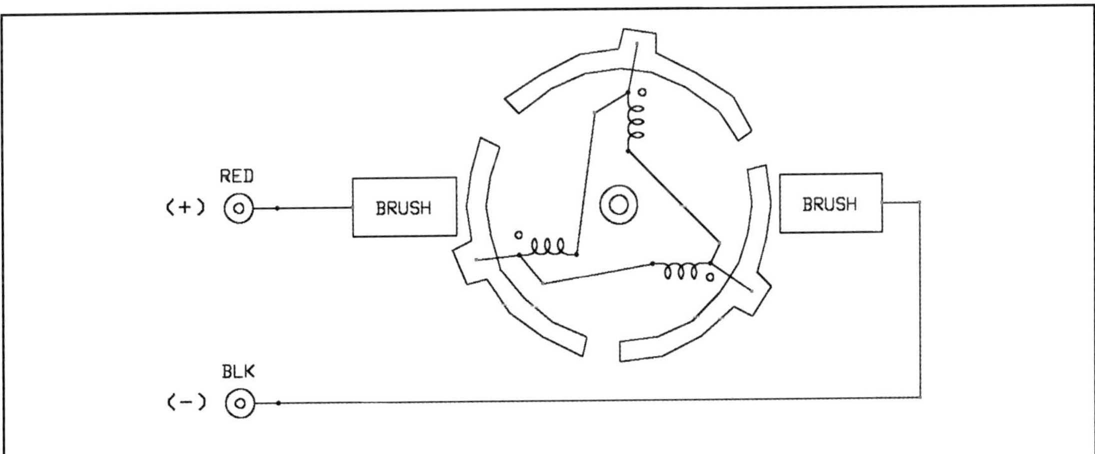

Figure 6-14: *The wiring for the Copper Queen is simple, but the coils must be connected properly with respect to their dotted terminals. Power is delivered to the commutator through a pair of carbon brushes.*

Motor Brushes

Successful motor brushes must have several attributes. To start off, they have to be reasonably good conductors. This is because they are the vehicle by which electricity is delivered to the spinning commutator. Any energy lost in the brushes is energy that is not available to the rotor to produce torque.

Next, they must be flexible and compliant. This is because the brushes must skip from segment to segment as the commutator turns. It's also possible that the commutator may not be perfectly round or run perfectly true, so this is another reason why brushes need to be able to "give" a little bit.

Third, the brushes must be tolerant of heat and mechanical wear. A brush resting on a spinning commutator amounts to two materials rubbing mercilessly against one another. A brush material that's too soft is likely to be ground away by the commutator. On the other hand, if it's too hard, it will eventually wear grooves in the commutator and damage it.

We know that a counter-EMF is produced when coils are de-energized and their magnetic fields collapse. This counter-EMF can produce arcing that is hot enough to erode and pit some brush materials. The brushes must be made of a material that will withstand this kind of abuse.

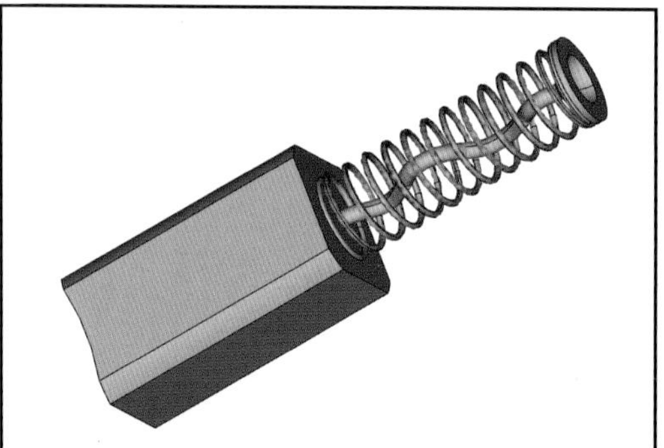

***Figure 6-15**: A typical motor brush consists of a block of graphite, a lead wire, and a spring (shown compressed here) which is used to maintain brush pressure on the commutator.*

Finally, the fact that the brushes rest on the commutator means they induce friction, which steals otherwise useful mechanical energy from the shaft of the motor. Brushes must be designed to impart as little friction on the commutator as possible.

A very common brush design involves the use of blocks of (carbon) graphite. The blocks, which are loaded into conductive metal guide tubes and backed with light springs, are oriented to force the graphite against the commutator. This type of design meets all of the requirements just discussed. Graphite is sufficiently conductive. Spring loading the graphite brush means it can be very compliant to irregularities in the shape of the commutator. Graphite is soft enough not to damage the commutator, but is tough enough to wear well. It can tolerate high heat, and because graphite is an inherently "slippery" material graphite brushes are, in essence, self-lubricating.

The brushes for the Copper Queen Motor were salvaged from some long-forgotten appliance. They feature graphite bodies, an attached wire lead (to improve conductivity between the brush and its holder,) and a captive spring that is used to load the graphite against the commutator. These types of brushes are not only common in junked motors, they can be purchased new as replacement parts at many hardware stores. See figure 6-15.

Each of the brush holders in the Copper Queen Motor were fashioned from a short length of brass tubing. The diameter of the tube was selected to complement the graphite brushes I had on hand.

One end of the tube was snipped repeatedly to produce small "petals," which were then overlapped and pounded flat to seal one end of the tube.

Next, a small square of brass sheet metal was drilled to fit onto the tube, and soldered into place to form a flange. Additional holes were drilled in the flange for mounting purposes. See figure 6-16.

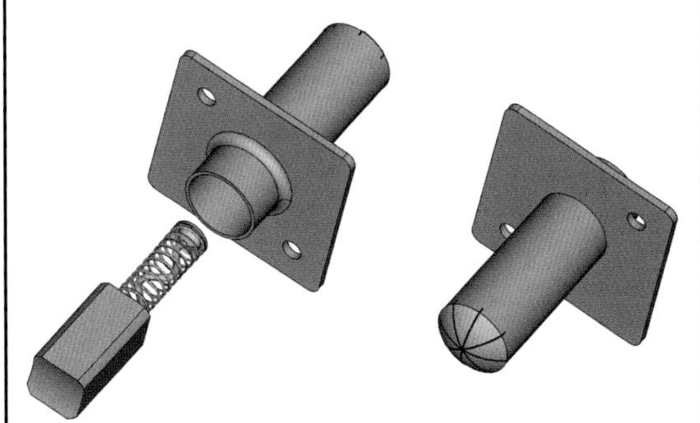

Figure 6-16: Two views of the Copper Queen's brush holders. On the left, a brush is about to be inserted into the holder. On the right, folded "petals" snipped into the end of the tube seal the tube and create a floor. The brush's spring engages this floor.

The brush holders were mounted in insulator blocks composed of nylon plastic, cut from a discarded kitchen cutting board. The insulator blocks measure 2 by 0.875-inch (⅞-inch), and are approximately 0.440-inch thick. Several holes were drilled into each block, one to accommodate the body of the brush holder, two for screws to secure the brush holder's flange to the insulator block, and two holes to be used by the suspension, the parts that hold the brushes in their proper position.

The suspension mechanism in the Copper Queen consists of two #8 threaded steel rods. The rods lie parallel to one another, and are spaced about 1-½ inches apart. They are mounted diagonally across the open mouth of the fruit basket, which comprises the motor's frame. Looking into the rear of the motor (the open mouth of the basket), the rods start at roughly the 11 o'clock position and cross to the 5 o'clock position.

The ends of the rods are rigidly attached to the rim of the basket using a variation of the sheet metal tabs or clamps described earlier.

Before the rods were attached to the motor housing, the insulator blocks were slid onto the rods, along with eight nuts and washers, which allow the insulator blocks to be located and anchored anywhere along the length of the rods. The relationship between the suspension, the insulator blocks, the brushes, and the commutator can be seen in figure 6-17.

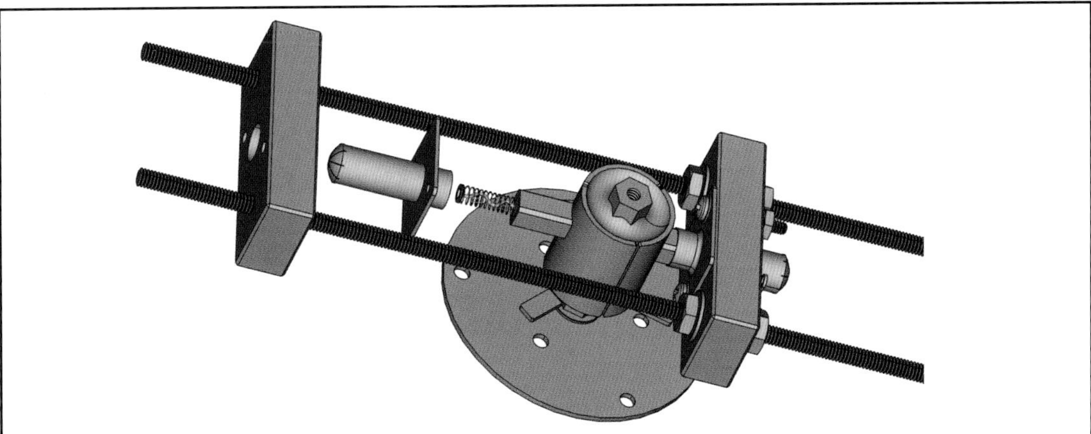

***Figure 6-17**: Motor brushes are inserted into brush holders and the brush holders are inserted into (and bolted to) insulator blocks. The insulator blocks are slid onto a pair of threaded rods and fixed into position with nuts. The threaded rods position the brushes over the commutator.*

Because the graphite brushes are spring loaded, they must be compressed and confined to their holders through manual means until assembly of the brush holder and suspension is complete. This can be accomplished with a bit of adhesive tape. Once each brush holder is in its correct position, the tape can be removed, and the spring will drive the brush against the commutator. The proper distance between the brush holder and commutator allows the brush to extend a bit from the end of the holder, but not so much that it runs the risk of springing all of the way out.

Field Magnets

The poles of the rotor turn inside of, and work against, a static magnetic field produced by so-called "field" magnets. In real motors, similar to the Copper Queen in design, field magnets can be implemented as a pair of always-on solenoids, or as a pair of permanent magnets. For simplicity's sake, I chose to use permanent magnets.

The field magnets in the Copper Queen, like so many of the motor parts in this book, were salvaged from discarded junk. I was out walking one day, trying to get some exercise, and discovered a pair of car-stereo loudspeakers lying beside a dumpster. The speakers' cones were torn and their frames were bent, but I knew that inside speakers of this type, one can often find large, ceramic, permanent magnets. The magnets I harvested in this case measured about 2-¼-inch in diameter, and ½-inch in thickness.

While the skeletal nature of the Copper Queen's frame makes for interesting viewing, it once again posed difficulties where mounting of the field magnets was concerned. I had originally envisioned suspending the field magnets from attachment points on the fruit basket, but gave up on this idea in favor of a simpler solution.

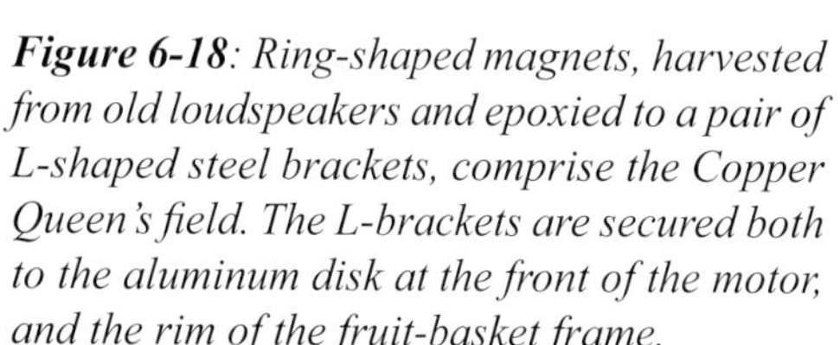
***Figure 6-18**: Ring-shaped magnets, harvested from old loudspeakers and epoxied to a pair of L-shaped steel brackets, comprise the Copper Queen's field. The L-brackets are secured both to the aluminum disk at the front of the motor, and the rim of the fruit-basket frame.*

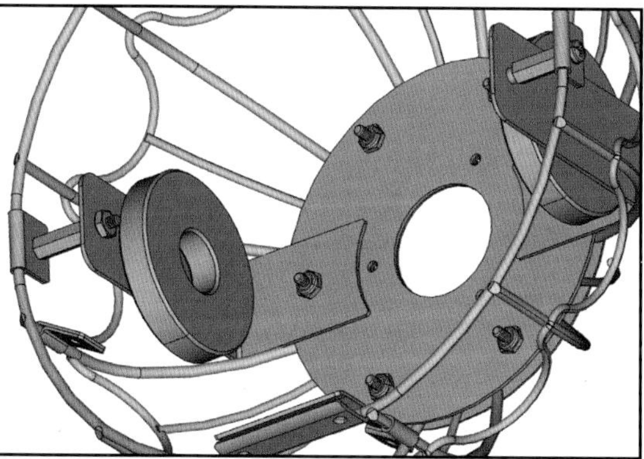

Most home improvement stores and lumber yards offer a selection of what is known as "hurricane brackets." Hurricane brackets, stamped from galvanized steel, can take the form of strips, angles, and even complex shapes. They're used to link wooden structural elements of a house together, so as to make that house more resistant to the damaging effects of high winds.

The support bracket for each field magnet began life as a strip of hurricane bracket, which I trimmed to length with tin snips, and then bent into an L-shape. I drilled a single mounting hole in the foot of each "L", sized to clear a #6 screw. The magnets were glued to the vertical section of the "L" with epoxy.

Some important details apply here. Before gluing the field magnets to the L-brackets, you have to identify the polarity of each field magnet. It's not necessary to know the absolute polarity of the magnets, but it is essential that the faces presented to the rotor be different. In other words, one should be a north pole, and one should be a south pole, but we don't care which is which. Second, make sure that the bracket is secured in some fashion during the gluing operation. If it's not, and you bring the magnet in proximity to the L-bracket, the bracket will jump up to meet the magnet. If it's wet with unset epoxy, you can imagine the mess this can create.

The foot of each bracket is attached to the flat metal disks at the front of the motor. It was not necessary to drill any additional holes in the plates, because there are already six screws and nuts present. I simply removed the nut from one of them, positioned the foot of a field magnet bracket over the screw, and replaced the nut. I repeated the process for the second field magnet. Looking into the rear of the motor (the open mouth of the basket), one magnet occupies a spot at roughly the 2 o'clock position, and the other lies at the 8 o'clock position.

When the motor is in operation, there is a substantial interaction between the rotor and field magnets. The gauge of the steel in the hurricane brackets I used was a bit on the thin side, which allowed the field magnets to wobble in and out. I stiffened them up by using a hex standoff, which is secured to the rim of the basket with a sheet metal "C" clamp. See figure 6-18.

Once my field magnets and brackets were built up and tested for a proper fit, I removed them from the motor, gave them a nice coat of hammer-tone paint, and reinstalled them in the motor.

Motor Timing

Timing is critical to the proper operation of any of the motors in this book, and the Copper Queen is no exception. The good news is that, if assembled in the manner I've just described, the motor should run without any initial adjustment. On the other hand, if you want to optimize the performance of the motor, maybe to increase its top speed or its low-voltage performance, this can be accomplished by changing the angular relationship between the brushes and the field magnets. This, in turn, can be accomplished in two different ways.

One way to effect a timing change is to detach the suspension mechanism (that holds the insulator blocks and brush holders in position) and rotate the points at which it's attached to the rim of the motor housing. As a starting point, the axis of the brushes should be perpendicular (90 degrees with respect) to the axis of the field magnets, but there may be benefit to increasing or decreasing this angular relationship.

Another way to effect a timing change is to loosen the screw that attaches the commutator to its mounting plate, rotate the commutator slightly, and re-tighten the screw. Rotating the commutator with respect to the rotor changes the time (and position) at which the rotor coils are energized and de-energized.

Shaft Arbor and Gear

Because the motor shaft on which the Copper Queen is based was harvested from a VCR, its length is limited. Only a short stub protrudes from the front of the motor. This length is more than adequate to allow it be attached to other machinery.

I regard the Copper Queen as a demonstration motor. It's purpose is to be observed and enjoyed, as opposed to being used to drive some real-world load. I therefore terminated the shaft based upon aesthetic — instead of engineering — considerations.

Rummaging through my junk box, I found a beautiful bronze gear that I decided would be a perfect accent for the Copper Queen Motor. The bore of the gear was somewhat larger than the Copper Queen's shaft, so I fabricated a simple arbor from a large bolt. The details are evident from figure 6-19. I attached the arbor to the Copper Queen's shaft with a set screw, slid the gear onto the arbor's shaft, and secured it with a nut.

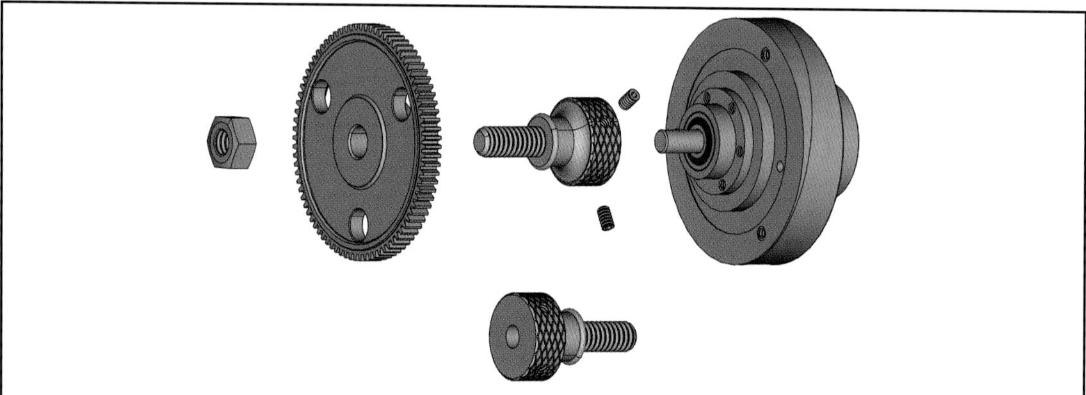

Figure 6-19: *A modified bolt is used as a shaft arbor to attach a decorative gear to the Copper Queen Motor. The detail shows the head of the bolt which was drilled out to accept the motor shaft, then drilled and tapped cross-wise to accept set screws.*

Needless to say, I have no intent to drive any loads with this gear. However, from an aesthetic standpoint, its size and composition blends well with the overall appearance of the motor. In addition, it's another component that's interesting to watch while in motion.

The Copper Queen is one of those constructions that can never translate fully to drawings or even photos. You have to see it to fully appreciate it. I've included figure 6-20 to show the heart of the motor — the rotor assembly — in its own glory.

Testing and Results

This motor is very novel in terms of the materials from which it's constructed, but the design and principles behind its operation are very traditional. The Copper Queen Motor is cousin to the electric motors that drive your power windows, start your automobile, drive your electric drill, blender, and vacuum cleaner, and vibrate your cell phone when the ringer is set to "silent".

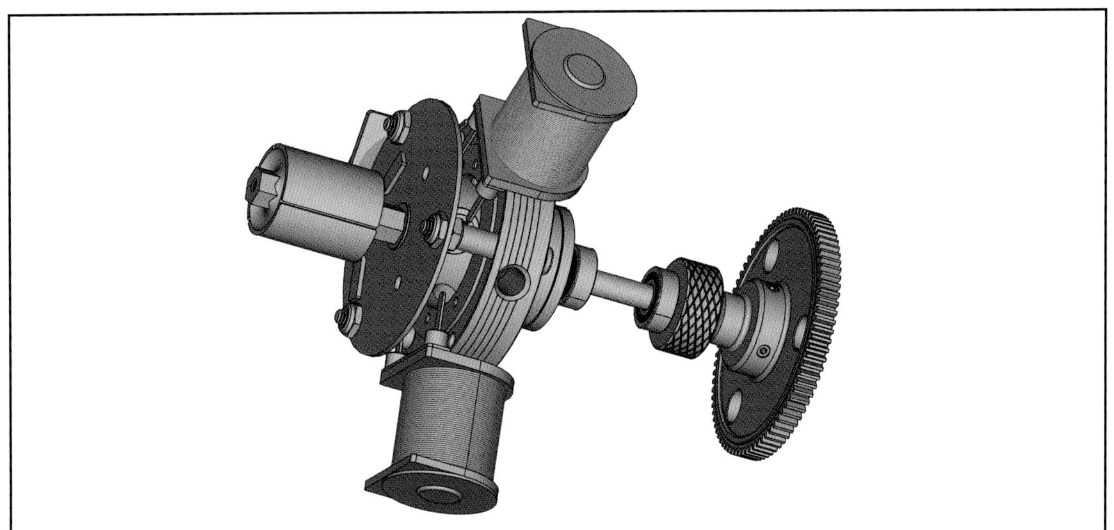

Figure 6-20*: A view of the Copper Queen's entire rotor assembly including coils, commutator, shaft, bearings, arbor and gear.*

Despite the crude nature of its construction, the Copper Queen Motor is unconditionally self-starting, smooth running, and relatively efficient. According to data I collected (see figure 6-21), its rotor will start on as little as 6V. At 29V, the rotor spins at nearly 2100 RPM, making it the fastest motor of those described in this book. It seems to be happiest around 20V or so, where rotation is brisk but smooth, and arcing at the brushes is not excessive.

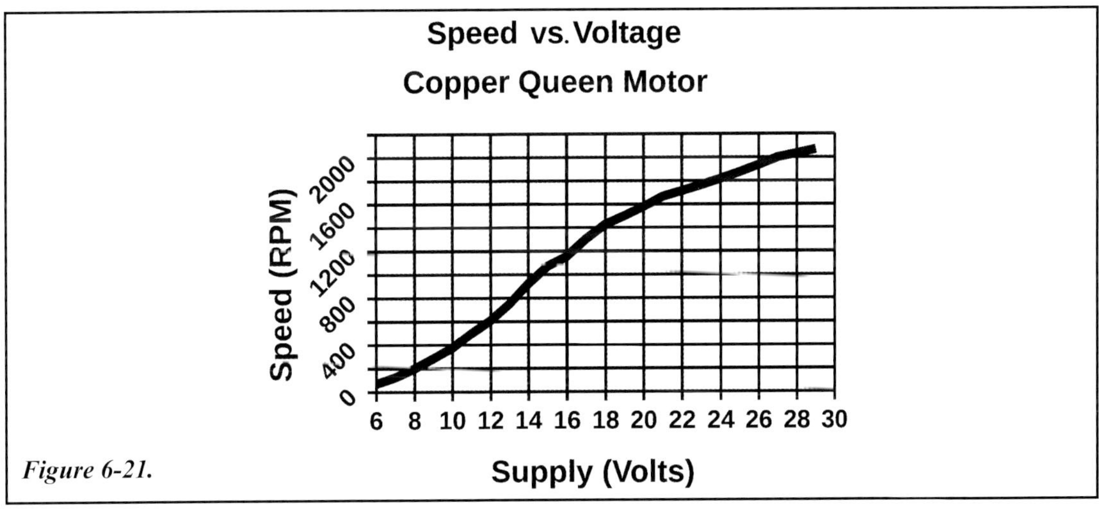

Figure 6-21.

Critical Considerations

Like the Twister Motor, there is little bad that can be said of the Copper Queen. But, it's always the case with projects of this nature that improvements can be made. The Copper Queen's rotor is reasonably well balanced, though additional effort could be expended to improve it. The solenoids I used are virtually identical, at least as far as their original intended purpose. However, an extra turn of wire on one versus the other translates to small differences in weight. When spun on the rotor at high speed, these differences appear as vibration.

In the case of salvaged rotor coils, like mine, balance could be improved by the addition of shim stock washers, or by drilling shallow dimples into the pole faces to remove metal. If you build a rotor using homemade coils, you actually have more latitude. In addition to the steps just mentioned, you have the option of removing magnet wire from an overly-heavy coil, to bring its mass on par with its twins.

The benefit to improving rotor balance extends beyond simple bragging-rights. Over the long haul, vibration can be damaging to bearings, resulting in reduced life.

This motor is also very crude with respect to the gap between the rotor and the field magnets. Reduce this gap, and the motor's torque will undoubtedly increase. Increased torque means a greater ability for the motor to do useful work, and may also result in increased operating speed.

The shape of the field magnets in the Copper Queen Motor is such that the opportunity for interaction between the rotor coils and field magnets is limited. First, the surface area of the face of each field magnet is comparatively small, and, because the field magnets were salvaged from loudspeakers, they are doughnut-shaped, with large holes in their centers. These characteristics work to degrade the performance of the motor.

A disk of soft iron plate could be placed upon the face of each of the field magnets to improve matters. A rectangular chunk of iron is even better, because the ends can be bent into a crescent shape. Together, pole faces like this assume a shape like a pair of parentheses, partially enclosing the rotor. This serves not only to distribute the flux of the field magnets where it is needed, but maximizes the time over which a given rotor pole can interact with a given field magnet.

A final thought: extreme care should be exercised when the Copper Queen is in operation. Any piece of homemade machinery, turning at high RPM's, has the potential to come apart and hurl components into the air. Wearing safety glasses while the motor is in operation is a good idea. Also, remember to keep fingers clear of the rotating parts.

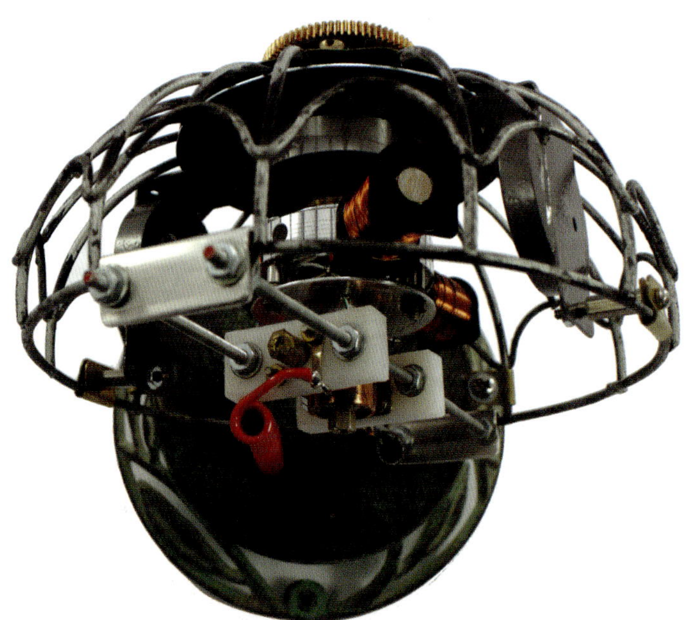

Chapter VII
Parting Thoughts

I may be biased, but I think that the preceding chapters more than make the case that interesting, attractive, and reasonably sophisticated model motors can be built with little more than simple hand tools. The only requirement is patience and the willingness to make creative use of recycled materials and re-purposed odds and ends. That point having been made, this would seem a logical place to conclude the book.

On the other hand, creative thought can't be switched off like a light switch, and insight does not necessary cease to emerge just because a manuscript has been completed. In the end, I decided to add a few pages to this book in order to share some parting thoughts with regard to technical resources, power supplies, instrumentation, and an experimental method to balance rotating parts electronically.

Technical Resources

It is always difficult, when writing a book of this kind, to properly scope the content. My intent is to appeal to the largest possible audience. Members of that audience may range from Ham-radio and electrical-engineer-types (with formal electronics knowledge), to inquisitive garage tinkers and eager weekend-machine-builders with little or no familiarity with electricity.

If you belong to the latter category, I think it a useful exercise to become better acquainted with what electricity is and how it behaves. A basic (but useful) understanding is within anyone's grasp and does not require use of overly complex mathematics. On numerous occasions I've been asked for advice on where to begin. Let me therefore offer the following information.

Every year, the American Radio Relay League publishes a new edition of the *The ARRL Handbook for Radio Communications* (formerly titled: *The ARRL Radio Amateur's Handbook*). While predictably slanted toward the topic of ham radio, the latest edition now contains six volumes packed with all kinds of useful electrical information. Among the relevant topics addressed in this set is an introduction to basic electrical theory, a discussion of practical design principles, tips on building real equipment, and a useful discussion of measurement instruments and how to use them. Frankly, the information in just one of those volumes is worth the price of the entire set. You can visit the ARRL on their web site at **www.arrl.org**.

Because a new edition of the handbook is released every year, there is always a substantial supply of older, used versions, that can be purchased at low cost. Check used book stores, hamfests, or the usual internet auction sites.

Another fine work that I routinely recommend is Horowitz and Hill's, *The Art of Electronics*. Granted, the book is overkill if all you plan to do is build a Peewee Motor.

On the other hand, if after digesting its introductory chapters you'd like to advance to understand how the transistor coil drivers in the Twister Motor really work, or how to build your own power supplies for the machines you build, it is a tome well worth the investment.

The Internet offers learning opportunity not even conceived of when I was young. Besides being a vehicle by which physical books, new or used, can be located and purchased, the Internet offers accessibility to countless volumes that are available for download for free. Among these are vintage books, no longer in print, that have been scanned and made available as PDF files by enthusiasts.

Speaking of free, an excellent resource, available in the form of downloadable PDF files, is the *Navy Electricity and Electronics Training Series*, created by the United States Navy. Twenty-four modules representing more than four thousand pages of material cover topics as diverse as matter and energy, microwaves, digital computers, magnetic recording, and fiber optics.

With a publishing date of 1998, it might be considered a bit long-in-the-tooth, but this has little bearing on the topics that apply to our Marvelous Magnetic Machines. Of particular relevance are Modules 1 through 4 (introductions to matter, energy, direct and alternating current, transformers, and wiring), Module 5 (introduction to generators and motors), and Module 21 (dealing with measurements).

In the appendix section of this book I have provided a list of books and some URLs. Some were used in the creation of this manuscript, some were merely referenced here. Some were not referenced but contain potentially useful information. Some feature home-built motor projects.

Resources no longer in print can sometimes be found in electronic form at the Internet Archive, Project Gutenberg, and elsewhere.

With an internet connection and a computer, the world is your oyster.

Power Sources

Most of the motors in this book will turn over and run when powered by an electrical supply in the 0-to-24V range. The amount of current demanded by motors of this size is equally modest, usually less than 1A. Suitable energy sources for voltages in this range fall into two broad categories: batteries, and mains-fed power supplies.

An ordinary flashlight cell of the "D" variety will provide 1.5V of electricity. By connecting numerous cells in series, as shown in figure 7-1, it becomes possible to deliver larger voltages. Eight cells in series, for example, gets you 12V. Sixteen cells in series tops out at 24V. By adding a "tap" switch to the battery pack, as shown in the schematic in figure 7-2, the output of the battery becomes adjustable.

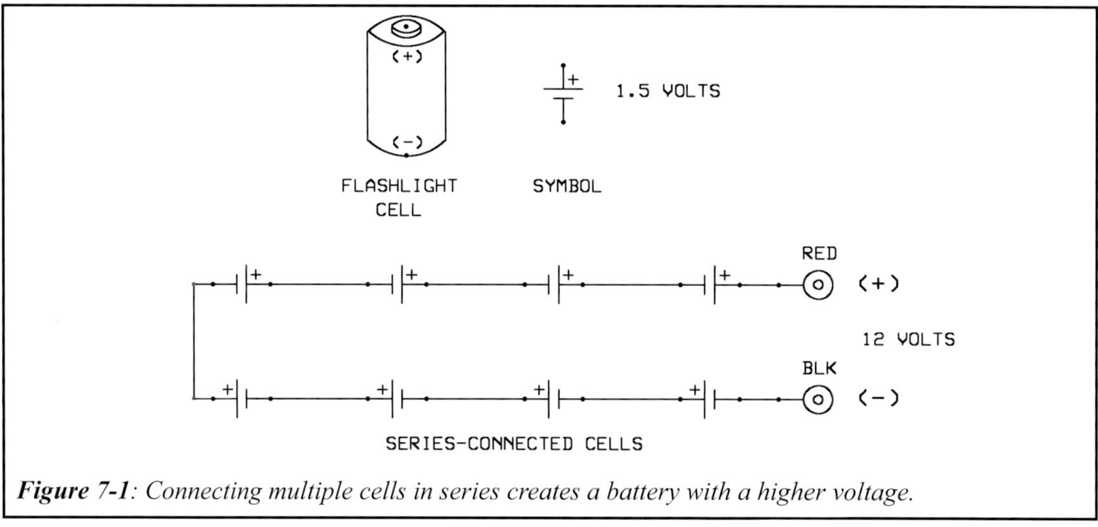

Figure 7-1: *Connecting multiple cells in series creates a battery with a higher voltage.*

How does one "connect" flashlight cells together? Commercial battery packs are linked together with nickel ribbon that is spot-welded to cell terminals. I've seen solder and wire used to the same effect, though I'd advise against it. The prolonged heat associated with soldering can damage flashlight batteries and even cause leakage or explosion. The best bet is to purchase battery holders from an electronics supply house. These are inert and can be safely soldered/wired as needed. Simply snap the cells in when it's time to play.

Ordinary flashlight cells are depleted as their energy is consumed, and eventually reach a point where they must be discarded. The use of battery holders makes replacement easy. However, power-hungry motor designs can drain batteries quickly, and recurrent replacement of dead cells becomes expensive over time.

An alternative is to make use of rechargeable cells, like certain alkaline cells, nickel-cadmium cells, or lithium-ion cells. The chemistry in each of these types of cells differs, as does the optimum (and safe) charging profiles. If rechargeable cells form the heart of your experimental power supply, it's probably best to use commercial charging equipment designed specifically for the type of cell you're using. This is particularly true for lithium batteries, which are prone to frightful and catastrophic failure if mis-handled.

Lead acid gel-cell and absorbed glass mat (AGM) batteries provide another option. From a chemistry standpoint, both are materially the same as your common automotive starter battery. However, while an automotive battery contains an acidic electrolyte that can be spilled if the battery is tipped to one side, the electrolyte in gel-cells and AGM batteries is sequestered and can't spill. Lead-acid batteries can provide significant amounts of current and will run small model motors for a long time before becoming discharged. The best way to refresh them is with a commercial trickle charger.

Gel cell and AGM type batteries can be purchased on-line, and at specialty battery stores. They may be available in certain toy stores, as they have been used as the power source for child-sized electric vehicles. Salvage opportunities for lead acid batteries include uninterruptable power supplies (UPS) for home computers and old alarm systems.

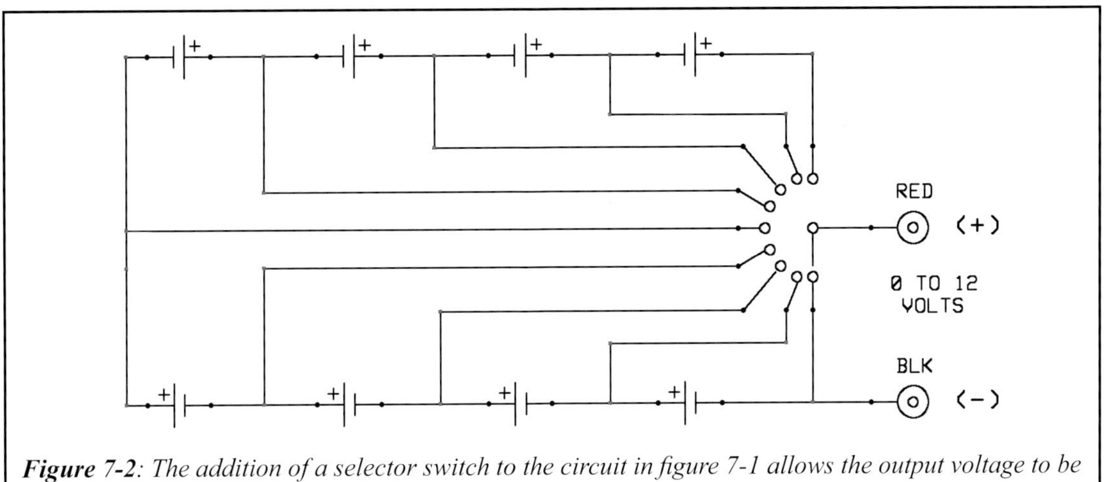

Figure 7-2: *The addition of a selector switch to the circuit in figure 7-1 allows the output voltage to be adjusted from 0 to 12V in 1.5V increments. The switch is shown here in the 0V (off) position.*

A battery that is too stale or too worn out for UPS use may still be perfectly serviceable as a power source for a small model motor.

While cells and batteries of the type just described represent a very low shock hazard, there still exists a potential for danger. A battery that has been short-circuited because of a wiring error, or because something metallic has crossed its terminals, may deliver large amounts of current, even if only for a brief period of time. High current produces heat, and that heat may result in the rupture of the battery casing, molten insulation and wire, and even fire. Exercise caution, particularly when using lead-acid batteries of any size. Wrists have been burned and even fingers have been amputated when metallic jewelry was accidentally laid across a high-power battery's terminals. Consider yourself warned — **one should not wear jewelry when handling batteries**.

Another option for powering motors is through the use of a power supply that converts the 120V AC present in a wall outlet to a lower, safer, DC voltage. So-called "bench-top" power supplies, designed and marketed specifically for laboratory work, are common and readily available. Used power supplies can be found at substantial savings on internet auction sites. While you're at it, check the Internet for information on local ham radio clubs. Most hams enjoy talking electronics, and may know where an old supply can be obtained for a few dollars.

Re-purposed commercial power supplies are another way to go. Literally millions of printers, laptop computers, video games, pieces of audio/visual gear, tape recorders, calculators, cell phones, and LCD monitors have been sold with accompanying "wall-warts". Wall-wart is a pejorative term for those black plastic boxes that plug into wall outlets, and provide low-voltage electricity for electronic devices.

Wall-warts seem to have that same mystical quality that socks in a dryer have. Sometimes they simply disappear, never to be found again. Other times, they reappear, though long after the appliance they once powered has been discarded. Thrift and charity resale shops usually have boxes full of them, and the tables at garage sales and church rummage sales always hold a few. Wall-warts represent an excellent source of relatively-safe power because they are ubiquitous, dirt-cheap, and inherently designed for consumer safety.

One of the more capable re-purposed power supplies commonly available are those that have been salvaged from discarded desktop computers. A typical ATX power supply, like the one lying in my junk box, is rated at 3.3 DC volts at up to 28A, 5 DC volts at 35A, and 12 DC volts at 16A.

Most supplies of this type feature an on/off switch at the rear of the supply, along with a three-terminal jack where the power cord is attached. At the opposite end, several multicolored cables emerge. The largest of these terminates in a 20 or 24-pin connector, that looks something like an harmonica. This is the plug that would ordinarily attach to the computer's motherboard. Other cables are terminated in smaller connectors that serve to power floppy, hard disk, and CD-ROM drives.

Often, wire colors on these devices follow a standard. Black wires represent ground or "common", for all of the voltages produced. These are typically interchangeable and can all be connected together as a group, if desired. Red wires usually deliver 5V. The orange wires supply 3.3V. A brown wire, if it appears in the bundle, is likely to be a sense line for the 3.3V output, and should be connected to one of the orange wires. The yellow wires coming from the supply are 12V lines.

In order to activate an ATX supply, it must be tricked into thinking that it's still residing inside of a computer. To accomplish this, the green wire on the "harmonica" connector must be wired to one of the black common wires. Once done, the supply can be turned on and off with the power switch at the back of the supply.

Computer power supplies belong to a class of electronics called "switched mode power supplies". It is a characteristic of this type of supply that it may not function properly unless a certain minimum load is always present. This can be guaranteed by wiring a 10Ω, 10W resistor across the 5V terminals of the supply, that is to say, between the red and black wires. An automobile tail lamp would probably work just as well.

A word of warning is in order. While I was able to locate an old PC power supply in my junk box and make productive use of it in the manner just described, supplies salvaged from different makes and models of computer may vary. Wire colors may vary. The Internet is teaming with reference information describing pin-outs and wire colors for all types of computer power supplies. It is always prudent to double-check one's information before proceeding.

And, like the lead-acid batteries mentioned earlier, power supplies of this type are capable of delivering large currents. I would not recommend the use of any high-current power source without the inclusion of fuses to protect the equipment (and the operator!) from accidental short-circuits.

Ideally, the power supply (or batteries, should you take that route) should be enclosed in an insulated box and fitted with external binding posts. Inside, a fuse should be wired between each terminal of the supply, and the corresponding binding post. By building the equipment this way, any circuit fault occurring outside of the box can have no effect worse than the destruction of one of the fuses. Computer power supplies need ventilation.

Don't forget to account for that. Finally, as you've no doubt gathered, I am a strong advocate of safety glasses when you tinker. Better safe than sorry.

It's possible to purchase suitable step-down transformers or extract them from discarded equipment to build a power supply from scratch. By combining a transformer with a rectifier bridge (to convert AC to DC), and an optional filter capacitor, a simple linear supply can be built that will power any of the motors described. A general plan for such a supply can be seen in the schematic in figure 7-3.

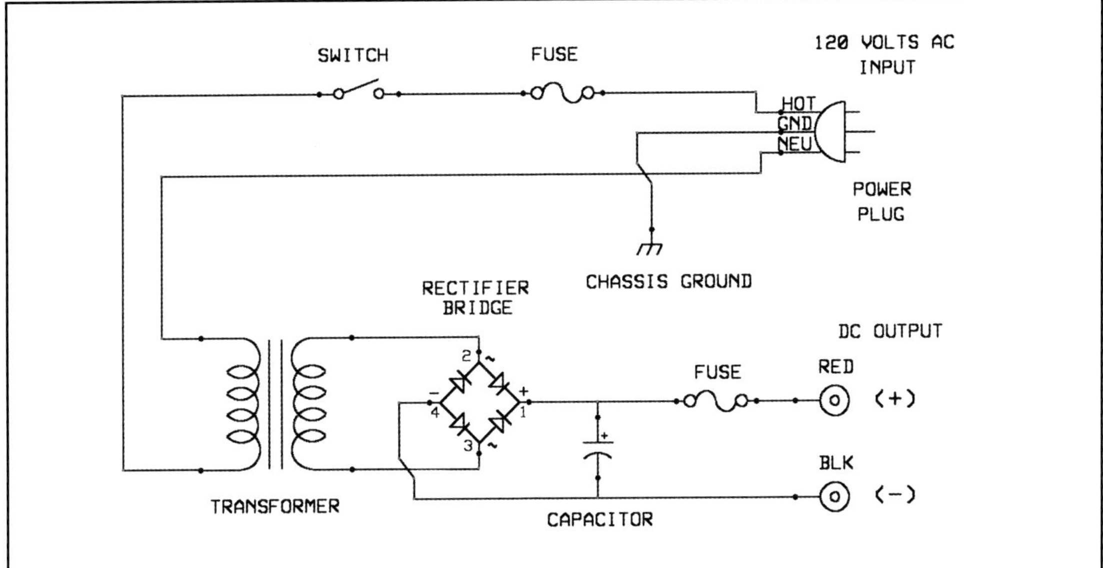

***Figure** 7-3: Schematic for a typical, unregulated, "linear" power supply, which converts household current to low-voltage DC. The circuit is simple enough to build, but this should only be attempted by those with sufficient knowledge of (and respect for) the dangers of working with wall-outlet power.*

The input or primary side of the transformer must be appropriate for the line voltage at your locale (typically 120V AC in the United States). House current is dangerous, so all of the wiring at the input side of such a supply should be switched, fused, and well-insulated. Ideally, the circuit should be constructed in a metal box, and that box itself grounded by connecting it to the green safety-ground wire of the power cord.

The secondary side of a usable transformer might be rated at 12 or 24V and must be capable of delivering the current your motors require (possibly up to several amps). Draw too much "juice," and you're likely to cook the transformer. A power supply intended for experimental use can therefore benefit from an additional fuse on the output-side of the circuit.

Rectifier bridges can be purchased or salvaged (old PC power supplies often contain nice rectifier bridges), but manufacturer data sheets should be consulted to make sure the bridge is capable of withstanding at least twice the transformer's output voltage. The bridge must also have the necessary current-carrying capacity. The operating voltage for filter capacitors should be at least twice what the supply can produce.

This is all well and good if you know your way around basic electronics. If not, re-purposing a wall-wart gets you the same results with far less concern.

Measurement Instruments

The choice and correct application of measurement instruments is a book-length topic in its own right. For those readers without an electrical or electronics background, I respectfully refer you back to my comments, earlier in this chapter, on technical resources.

For the moment, I'll assume that you have a sufficient grasp of basic electricity and the operation of the motors described in this book to appreciate the value and utility of some basic measurement equipment.

Multimeters

No one who is tinkering with electricity should be without a multimeter. By the term "multimeter," what I am referring to is an all-in-one measurement instrument that allows you to read voltage, current, and resistance.

Many older multimeters are analog in nature, which is to say that they are based upon a meter movement in which a pointer moves against a graduated scale. If you should happen to inherit one, or stumble across one at a flea market, garage sale, or internet site, try to curb any prejudice, and give it fair consideration. While analog instruments predate their modern digital descendants, they are more than adequate for use with projects like those in this book. If you get your hands on one of the classics manufactured by Simpson or Triplett, you should consider yourself blessed. For some measurement applications, I actually prefer an analog multimeter to a digital one.

Digital multimeters, or DMMs, can be found everywhere and at a wide range of price-points. I've seen Chinese imports sell new for as little as five bucks. My experience has shown that in many cases, ultra-ultra cheap devices tend to be garbage, but your mileage my vary.

To avoid damaging a multimeter, it is important to set it properly for the type of measurement to be made, and range it properly for the magnitude of the signal expected. Generally there are switches or buttons on a multimeter that allow the operator to select what they intend to measure (i.e., volts, amps, ohms, etc.) The same control may be used to set the intended range of the instrument. Newer/better instruments tend to "auto-range," that is to say, the instrument determines and sets the input range by itself.

Figure 7-4 summarizes the manner in which a voltmeter is used. It's always connected across a power source or load, and is use to measure electric potential, like the voltage present on the terminals of a battery, for example. With analog meters, one must exercise some caution with regard to polarity when applying the meter's test leads. The red lead goes to positive, the black to negative. Analog voltmeters show positive values only, so if you accidentally swap the test leads, the meter needle will simply bang against its lower stop and sit there. DMMs are more forgiving in this regard. If the leads on a DMM are backwards, the display will show a minus sign in front of the reported value.

Electrical current, measured in amperes, represents flow. An ammeter must therefore be placed in series with a circuit. See figure 7-5. Never connect an ammeter directly across the terminals of a battery or power supply, because the meter will act as a dead short. If you're lucky, you'll blow a replaceable fuse inside of the multimeter's case. If you're not lucky, you can burn the meter beyond repair.

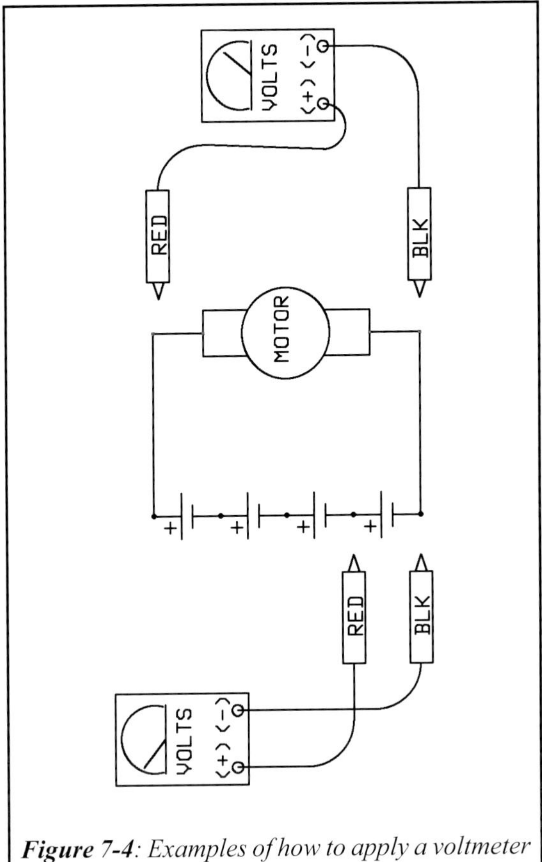

***Figure 7-4**: Examples of how to apply a voltmeter to a simple circuit comprised of a battery and motor. The voltmeter to the left measures the potential on a single cell of the battery. The voltmeter on the right measures the voltage applied across the motor terminals.*

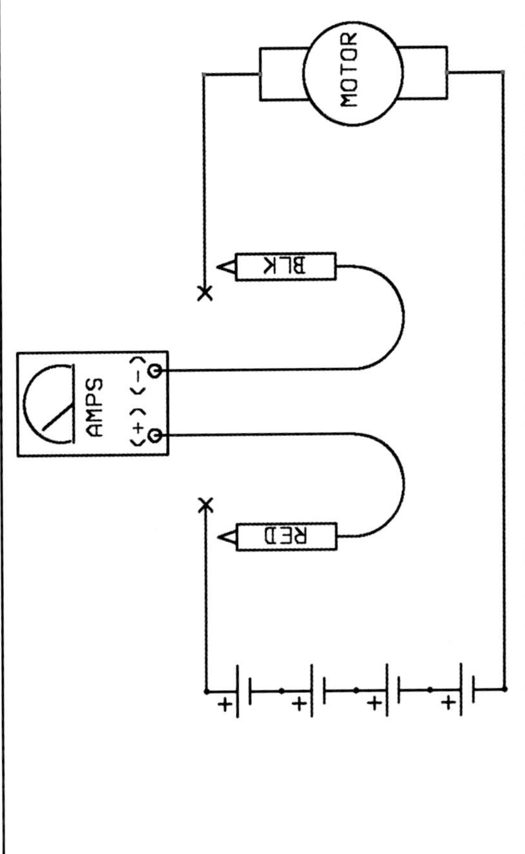

***Figure 7-5**: An examples of how to apply an ammeter to a simple circuit comprised of a battery and motor. The circuit loop is first broken and the meter is then inserted in series.*

The Ohm scale on a multimeter measures electrical resistance. Never attempt to make a resistance reading on something that is energized. More likely than not, you'll end up destroying your meter.

Analog multimeters usually have a switch with which one can select a resistance range. Use a low range (tens of Ω) if you're checking a switch for continuity, or testing the windings in a solenoid. Higher ranges (hundreds or thousands of Ω) are appropriate if you're measuring large-value resistors in electronic circuitry. As mentioned earlier, better digital meters can auto-range.

Some DMMs, like my Fluke, contain a beeper which chirps to verify continuity. This is handy, because I don't need to raise my eyes from my work to verify a connection or confirm that a leaf switch, for example, has closed.

Tachometers

Since motors are rotating machinery, it is natural that one might be interested in the rotational speed of the moving parts. Measuring the speed of a rotating shaft is usually done with an instrument called a tachometer, and these come in several varieties.

Contact tachometers typically feature a short rubber-tipped shaft that protrudes from the side of the instrument. The rubber tip is placed in contact with the rotating part you wish to measure, and an analog meter or digital display will tell you the measured rotations per minute. A quick search on the Internet reveals that these can be purchased new for as little as fifty dollars.

Optical tachometers measure rotational speed using light. Before making the measurement, one marks a flywheel or other rotating part with a strip of reflective tape. A light source inside of the instrument casts a beam on the rotating part, and a reflection occurs every time the tape intersects the light beam. A counting circuit in the instrument measures the rate of these reflective pulses and indicates shaft speed on a display. With a little bit of web surfing, new instruments of this type can be purchased for less than twenty dollars.

Frequency Counters

An alternative to the tachometer is the frequency counter. This instrument counts a stream of incoming electrical waves or pulses, and divides that count by elapsed time as tracked on an internal clock. The result is a number expressed in units called "hertz" (Hz). One Hz is the same thing as one pulse per second. I mention frequency counters as an alternative to tachometers because used counters are cheap, and are general-purpose devices that can be used for other electronic applications. Some models of DMM, by the way, have a frequency counter option built right in, eliminating the need to purchase additional equipment.

Since the motors described in this book are electrical in nature, and all of them involve timing, certain electrical events must occur every time a flywheel or rotor turns. In an attraction motor like the Peewee, for example, we know that the leaf switch which fires the solenoid does so once for every revolution of the flywheel. If we connect a frequency counter to the solenoid, the counter will display, in effect, the rotational speed of the flywheel in Hz.

This leads me to an important mathematical distinction. It is usual to express shaft speed in terms of RPM (rotations per minute). The frequency counter displays Hz (pulses per second), which in the context of the Peewee means "rotations per second". If you are interested in the RPM, it is necessary to multiply whatever appears on the frequency counter's display by 60 to make the conversion.

Other mathematical corrections may be necessary, depending upon the motor. For instance, in the case of the Twister Motor, we can use a frequency counter to measure the rate of pulses coming out of one of the Hall effect sensors. Recall, however, that the rotor on the Twister has three permanent magnets. So, for every revolution of the rotor, a given sensor is triggered three times. Stated another way, the frequency of pulses coming out of a Hall effect sensor on that motor is three times the speed of the rotor. To measure the Twister Motor's speed in RPM with a frequency counter, one must first note the number on the counter display, divide that number by 3 (to account for the multiple magnets), and then multiply the result by 60 (to convert Hz to RPM).

Oscilloscopes

One of the most useful instruments an experimenter can own is an oscilloscope. For those unfamiliar with this device, it looks vaguely like a television set. Superimposed on the screen is a ruler-like grid, which is used to measure the figures that appear on the screen. An internal "clock" causes a dot of light to sweep across the display, horizontally, at regular intervals. An input signal can deflect the dot up or down, in a vertical direction. The oscilloscope is a device used to create live graphs that plot changes in an input signal (Y axis) with respect to time (X axis.)

The 'scope's internal clock makes it capable of measuring motor speed. All we have to do is connect the input of the scope to monitor an electrical event in the motor. Using the Peewee Motor as an example again, the closure of the leaf switch will produce a series of pulses on the scope's screen. The first step in determining motor speed is to measure the horizontal distance from the start of one pulse to the next, using the measurement grid. Let's consider the hypothetical trace in figure 7-6.

We can see that the horizontal distance from the start of one pulse to the start of the next is about 2.4 squares, as measured by the grid. These squares are called divisions.

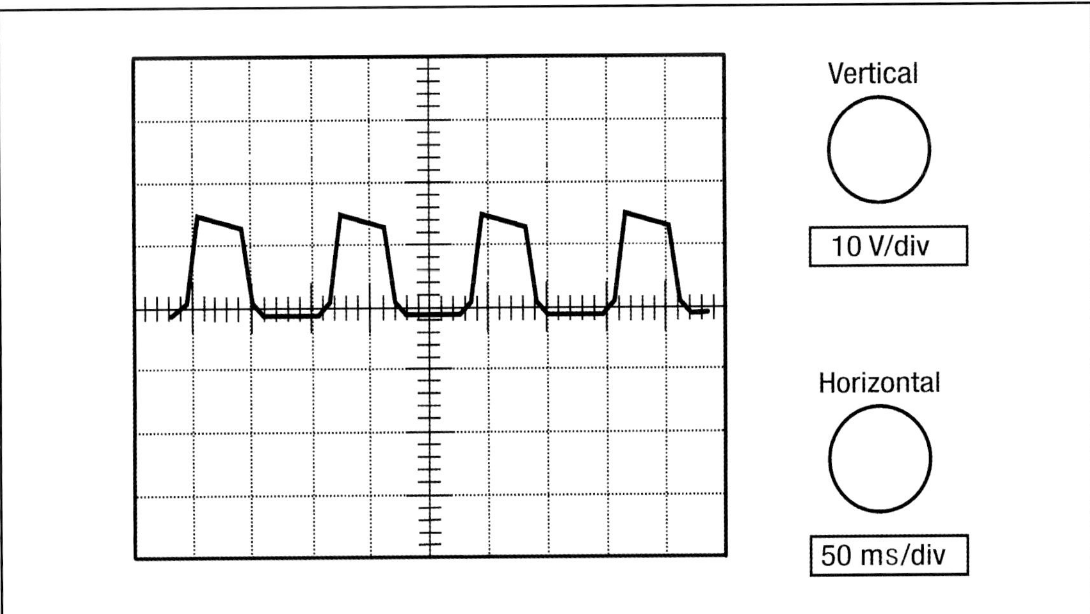

***Figure 7-6**: Using an oscilloscope to measure motor speed. The ruler-like scale on the 'scope screen indicates that the pattern repeats every 2.4 divisions. This measurement multiplied by the setting on the horizontal scale control (50 milliseconds per division) means each event takes 0.12s to complete. From this, frequency and ultimately RPM can be derived.*

All 'scopes feature a control that sets the speed at which the dot sweeps across the screen, which in turn establishes the numerical "weight" of each square on the screen's grid. For the purposes of this discussion, we'll assume the time-base dial is set to 50 milliseconds per division. This means that along the horizontal axis, each square represents 50 thousandths of a second of elapsed time. Therefore, the total time from one leaf switch closure to the next is 2.4 divisions (read from the measurement grid) times 50 milliseconds (ms) per division (as set by the horizontal control), or 0.12 seconds (s) per event.

We don't want seconds-per-event, what we're after is events-per-second (Hz), so this number must be inverted. This is done by dividing 1 by the 0.12s value. The result of this calculation is 8.33Hz.

Since, by convention, we'd prefer motor speed expressed in RPM, the Hz value must then be multiplied by 60. The Peewee Motor, in this case, is turning at 8.33 times 60, or 500 RPM.

Another control adjusts the vertical sensitivity of the instrument. The distance between the bottom and the top of the waveform in figure 7-6 is about 1.6 divisions. Based upon the vertical gain setting, each division represents 10V. Ten x 1.6 yields a pulse height of 16V.

There is insufficient space here to fully describe the utility and power of an oscilloscope, and the few paragraphs here simply can't do the topic justice. *The Art of Electronics*, mentioned earlier, contains in its "Appendix A" an introduction to the oscilloscope which, though concise, is very useful. Educational institutions and prominent manufacturers of oscilloscopes have made instructional materials available for free on the Internet. Tektronix's 44-page booklet, *Primer XYZs of Oscilloscopes* is one such example. An internet search will reveal many more.

Personal Computer-Based Instrumentation

The heart of many digital measurement instruments is an electronic system of components collectively referred to as an analog-to-digital (A/D) converter. The A/D looks at an input voltage level, for example, and then quantizes it. Stated more simply, it generates a number that represents the magnitude of the input signal at that particular instant. In the case of the digital multimeter, which we discussed earlier, a voltage applied to the input of the instrument is converted to a number representing the measured voltage, which is then posted on the instrument's display screen.

Success in making any measurement lies in the selection of an appropriate measuring device. Consider: an odometer, a ruler, and a micrometer are all instruments capable of making linear (length) measurements. However, an odometer which can resolve distance to one-tenth of a mile is no good for measuring the length of a wooden plank. A ruler or tape measure might be appropriate for measuring the plank, but are similarly useless for measuring the diameter of a human hair. My point is that different measuring instruments offer differing degrees of resolution.

The same is true for analog-to-digital converters. Resolution is often expressed in “bits”, which indicates the number of binary digits that the converter can express. A simple 4-bit converter can resolve no better than one part in sixteen (two to the fourth power). If such a converter were used in a digital voltmeter scaled to measure 100V signals, the resolution of that voltmeter would be no better than one-sixteenth of 100V, or 6.25V. A meter with this coarse a resolution would be useless for measuring the voltage of a 1.5V flashlight battery, for example.

A 16-bit converter, on the other hand, can resolve values to one part in 65,536 (two the 16th power). The same 100V meter described above, based upon a 16-bit converter, could resolve voltage differences as small as 0.0015V. Roughly speaking, the 16-bit converter can resolve signals with four-thousand times greater detail. There are factors that degrade real-world resolution from these theoretical values, but at the very least, you can see that more elaborate A/Ds are required if high-resolution measurement is desired.

Another defining attribute of an A/D is its conversion time. The process of capturing a snapshot of an input signal, and then converting that snapshot into a number, takes a certain amount of time. Some A/Ds can perform this conversion very quickly, while others may take a comparatively long time. High conversion speed, like high resolution, usually requires a more sophisticated converter, which can mean higher cost. The quality of an instrument’s analog-to-digital converter, and associated circuitry, has a lot to do with that instrument’s price tag.

As of 2016 (the last year I saw data for this), 9 of 10 households in the United States had at least one personal computer. Based upon the number of unwanted computers I’ve literally been given over the years, I’d guess that number to be even higher. My reason for mentioning this is that pretty much any personal computer built in the last twenty years has a sound port — a microphone-level or line-level input — with the ability to record audio signals. This capability depends upon an internal analog-to-digital converter, a component that can be pressed into service to make certain electrical measurements in the home laboratory.

Yes, special software is required to accomplish this, but much of it is available for free on the Internet. Try a Web search using the phrase “sound card oscilloscope” or “sound card frequency counter”.

Now, before you run off to take advantage of this, there are some pitfalls that you must be aware of.

First, to make use of the software, one must fabricate an interface cable to properly couple the computer’s sound port to the signals you wish to monitor or measure.

The physical connection to a typical personal computer is a jack which accepts a 3.5-millimeter stereo plug. The “stereo” designation refers to a plug with three conductors. These conductors include the point of the plug (referred to as the “tip”), a short metal segment adjacent the tip called the “ring”, and the larger conductive body of the plug (referred to as the “sleeve”).

When inserted into the computer's input jack, the tip links to the left audio channel, the ring connects to the right channel, and the sleeve acts as a common connection or "ground" for both. Note that this wiring arrangement is typical, but not universal. Certain laptop computers, for example, use 4-conductor plugs that provide both signal input and output through a common jack.

Computer sound ports are not designed for the direct injection of motor-level voltages, and doing so may result in permanent damage to the sound port, to the computer, or both. Circuit trickery can be employed to minimize this risk (we'll explore a few circuits in just a moment) but consider yourself forewarned. Given the ubiquity of old and unwanted computers, I'd recommend limiting experiments to computer equipment that you can afford to damage or destroy.

Finally, bear in mind that the sound circuitry in a typical computer has been designed for just that... sound. A signal is usually regarded as being an "audio" signal if it is an alternating current with a frequency in the range of 20 to 20,000Hz. Consequently, the range of frequencies over which computer sound hardware is usable is no wider (and usually much less) than this range.

Therefore, a computer's sound port cannot be relied upon to measure direct current (which is in effect 0Hz), so you can't use it to measure the voltage of a battery or direct-current power supply.

Input Adapters

One way to capture an input signal to the sound card is through inductive coupling. A small coil of wire, when brought near an operating motor, is likely to pick up magnetic disturbances that occur when the motor is in operation. Pickup coils, complete with a cable terminated with an audio plug, are commercially available and have been sold for use in recording telephone conversations. The beauty of inductive coupling is that no direct electrical connection is made between the device under test (your motor, for example) and your computer.

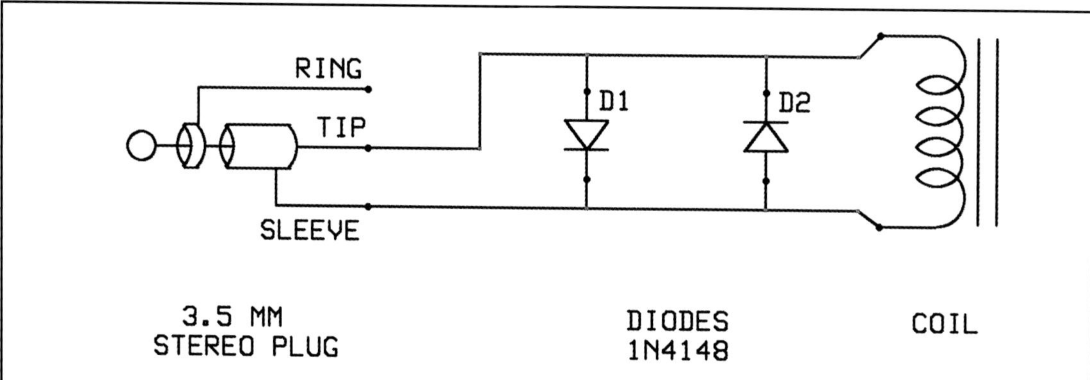

***Figure** 7-7: A pickup coil, connected to the line or microphone input of a computer, may detect magnetic field changes around an operating motor. Free software can be used to record and measure the detected signals. Diodes are employed to protect the computer input from excessive voltages caused by over-stimulation of the coil.*

You can buy such an adapter, or experiment with the design in figure 7-7. Assuming your computer uses the most common input wiring, your computer should "hear" the signals detected by the coil on the left audio input channel.

The output of the pickup coil is dependent on the magnetic field in which it is immersed, and if aggressively stimulated, could send undesirably high voltages into the computer's sound port. To protect against this, diodes D1 and D2 have been added. Wired as shown, they act as a limiting device to "clamp," whatever voltage the coil produces to a reasonable level, something on the order of 1V or less.

A similar circuit, but one that allows direct application of small electrical signals employs an audio transformer. Such transformers can be purchased new, or salvaged from junked transistor radios. The circuit in figure 7-8 allows direct electrical connection to the device being monitored while maintaining a protective electrical isolation between the device and your computer. Again, diodes are employed to enforce a limit on the amplitude of the signal seen by the computer.

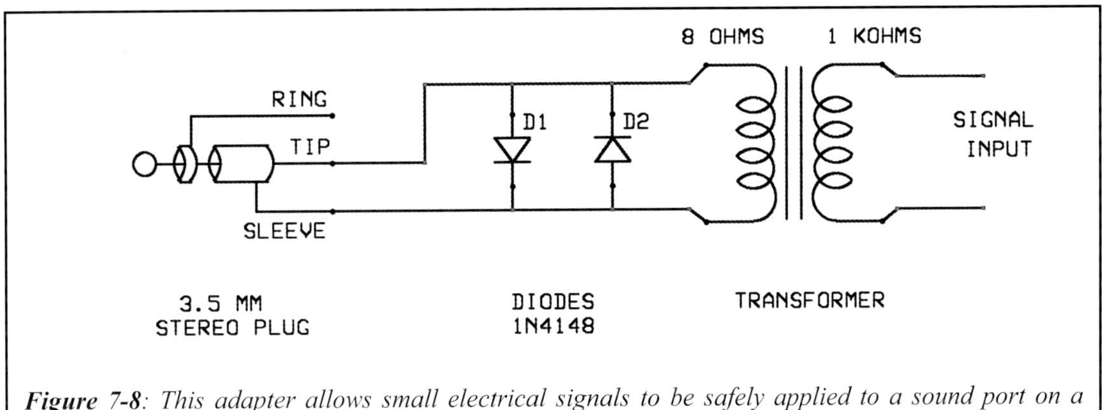

***Figure** 7-8: This adapter allows small electrical signals to be safely applied to a sound port on a computer. The transformer provides electrical isolation, the diodes limit the amplitude.*

Still another way to couple the computer's input to a motor being tested is with light. This is easily done with a part called an optocoupler.

Optocouplers come in a variety of flavors, but one of the most common forms looks like the device in figure 7-9. The part is vaguely U-shaped. One leg of the U contains an infra-red LED (light-emitting-diode) which acts as a source of light. The other leg of the U contains a phototransistor, a device whose conductivity changes upon exposure to light. If the LED is lit, light crosses the gap between the legs of the U and strikes the phototransistor. In the illuminated state, the phototransistor's resistance is low. If an object is inserted into the space between the legs of the U, the infra-red light from the LED is blocked. It can no longer stimulate the phototransistor, so the phototransistor's resistance rises significantly.

Now imagine attaching a little tab or "flag", composed of a piece of black electrical tape, to the shaft of one of our motors. If we position the optocoupler so that the flag can fly through the gap between the legs of the U then, as the motor rotates, light in the gap of the optocoupler will be blocked once for every revolution of the shaft. The phototransistor will detect this, and the resulting signal can be fed to our sound port.

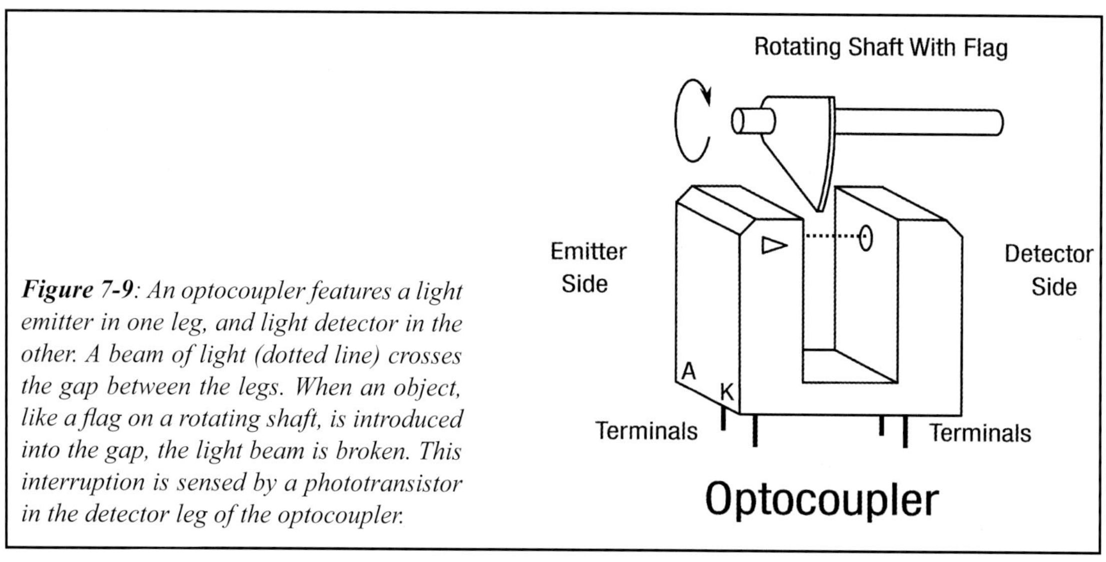

Figure 7-9*: An optocoupler features a light emitter in one leg, and light detector in the other. A beam of light (dotted line) crosses the gap between the legs. When an object, like a flag on a rotating shaft, is introduced into the gap, the light beam is broken. This interruption is sensed by a phototransistor in the detector leg of the optocoupler.*

Figure 7-10 is a schematic which represents the idea I just described. A 3V battery, comprised of two AAA cells in series, lights the LED through a 100Ω current limiting resistor. The battery voltage is also applied to the phototransistor through a 270Ω resistor. The output of the detector passes through a 1µF (microfarad) capacitor, which is included to isolate and protect the sound port from the direct application of battery voltage.

Where can you find an optocoupler? The answer is — everywhere. You can purchase them new through electronic supply houses for a dollar or less, on-line and elsewhere, though I have to add that I've never torn apart an old printer or fax machine and failed to find at least one. The key to making successful use of a salvaged optocoupler is to decode the cryptic markings often found on its case.

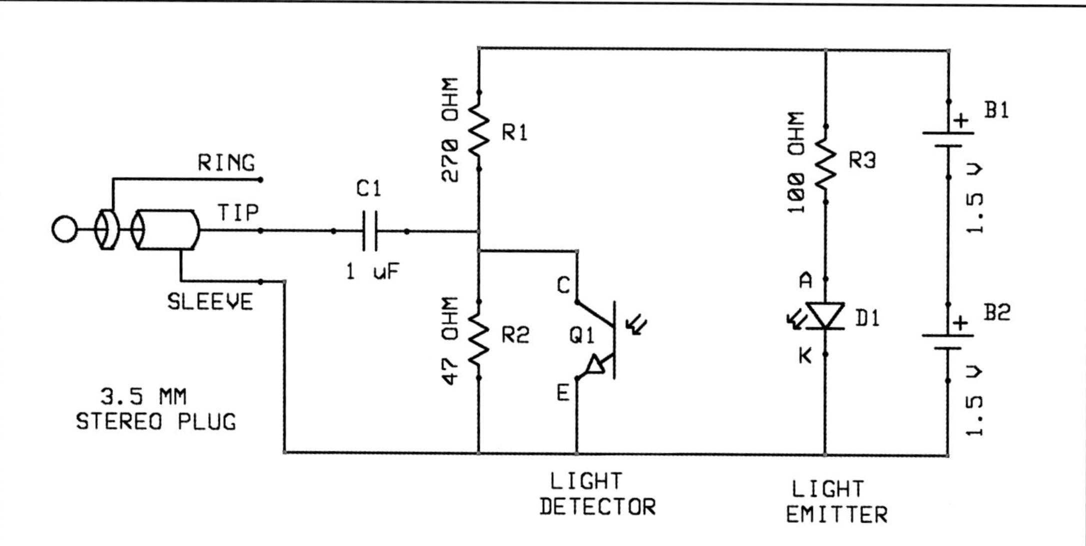

Figure 7-10*: Light-emitting diode D1 casts a beam of light on the photodetector Q1. Periodic interruption of that beam (caused by a moving shutter or "flag" mounted to a motor shaft) will turn Q1 on and off, generating a signal that will appear on the tip of the stereo plug. From this signal, with suitable software, a computer can infer the speed of the motor.*

Arrows pointing into or out of the detection gap will give you an idea which half of the coupler contains the LED, and which half contains the phototransistor. An arrow pointing into the gap signifies that leg as the one containing the LED emitter. An arrow pointing away from the gap signifies that leg as containing the phototransistor. If you see the letters “A” and “K” near the terminals, this designates the anode and cathode of the LED (refer to 7-10). If you see the letters “E” and “C” near some terminals, this surely designates the emitter and collector of the phototransistor. Note that polarity is important, and if you connect either the LED or phototransistor backwards, the circuit won’t work.

Here’s a quick tip for verifying that your LED is working; use a cell-phone camera to photograph the gap between the legs of the U while the optocoupler is energized. Most digital cameras have some sensitivity to infrared light, so while your eye will not detect a glowing infra-red LED, your digital camera surely will.

One of my favorite pieces of free, open-source, computer software is called *Audacity*. *Audacity* is intended as an audio recorder/editor, but because it displays waveforms against a time base (in a sense, like the oscilloscope), it can likewise be used to measure timing.

Figure 7-11 shows an example waveform that I captured on my laptop using an optocoupler in conjunction with *Audacity*. As you would expect, the signal is regular and repetitive — one pulse per revolution of the shaft I was monitoring. The horizontal distance between subsequent peaks, represents the time it took to complete one revolution. The time value can be measured using the scale, located directly above the waveform display. Alternately, you can drag your mouse to highlight a start-to-finish segment of the waveform, and read the start and stop times from the timing fields below the waveform display.

The horizontal distance between peaks in 7-11 is about 0.11s. Inverting this value (dividing it into one) gives you 9.09Hz. Multiplying that times 60 yields a shaft speed of 545 RPM.

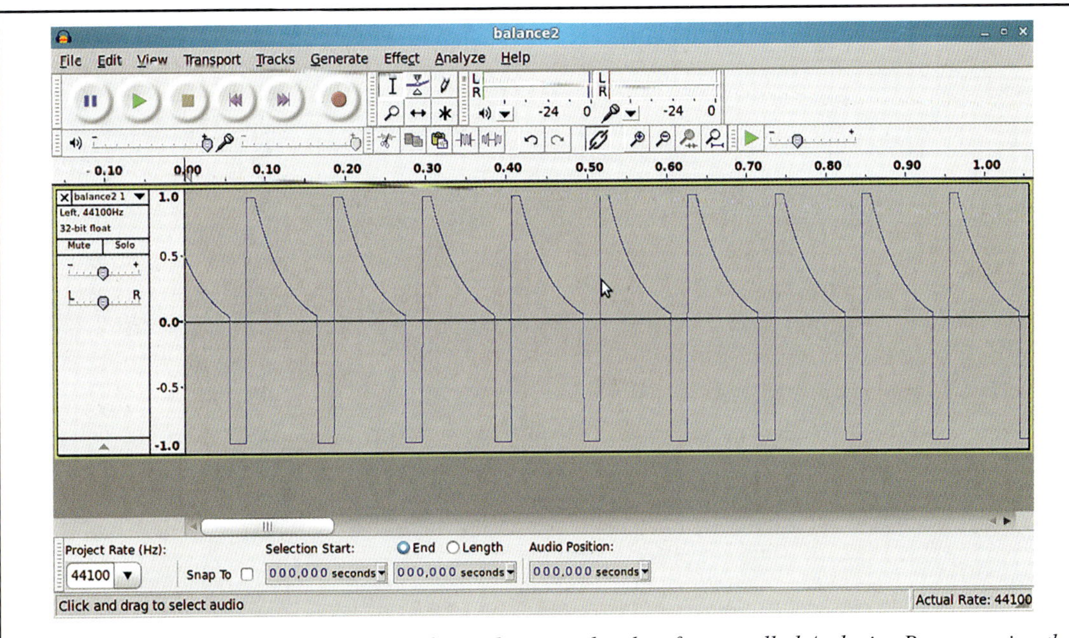

Figure 7-11: *A sample optocoupler speed signal captured with software called Audacity. By measuring the time elapsed between subsequent peaks, we can compute the speed of the motor that generated this plot.*

Thoughts on Electronic Motor Balancing

Most of the motors I've built show some evidence of vibration when they're in motion. This is particularly evident in the reciprocating attraction motors where a crankshaft is inherently unbalanced by the attachment of a connecting rod. In theory, much of this vibration can be eliminated by the addition of appropriate counterweights to the crank, flywheel, or other rotating parts. The trick is in knowing where to add the counterweights, and how heavy to make them.

The simplest way to locate and size counterweights is to balance the motor statically. This involves fiddling with the rotating parts of a motor to see if they "favor", a particular orientation, due to the action of gravity. For example, the flywheel in figure 7-12 is non-uniform, such that the mass in the A-half is slighter larger than the mass of the B-half. This flywheel will have the predictable tendency to come to rest with the A-half of the wheel in the downward position.

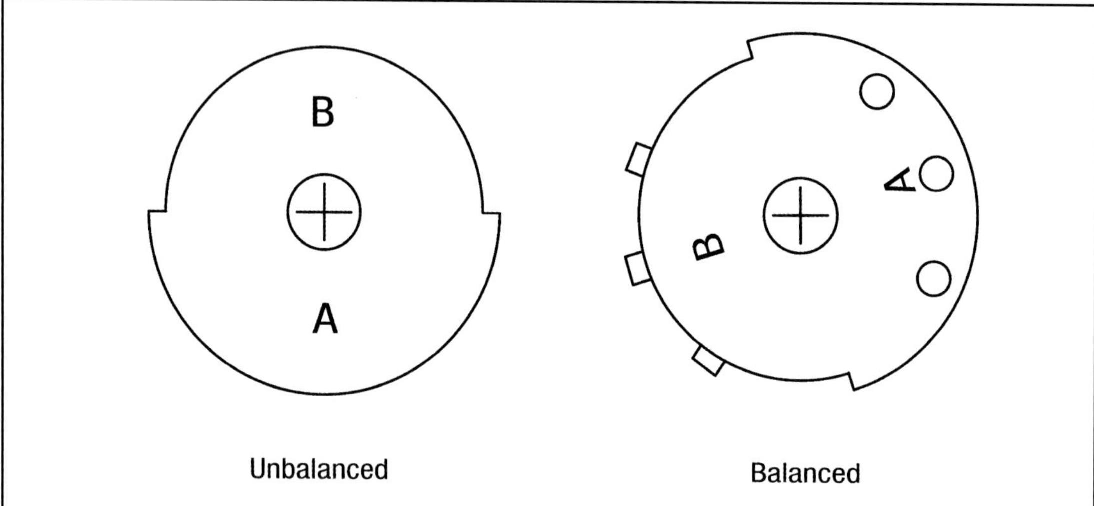

***Figure 7-12**: When in motion, an unbalanced wheel (left) is a source of undesirable vibration. At rest, it assumes a predictable heavy-side-downward orientation. Luckily, even grossly asymmetric bodies can be brought into balance through the redistribution of mass. Holes can be drilled into the A-half of the wheel to reduce the concentration of mass there. Counterweights can be added to the periphery of the B-half of the wheel to increase the mass there. The modified wheel (right) is in balance and shows no particular preference for orientation when it comes to rest. It will run with minimal vibration.*

Balancing the wheel fix requires a redistribution of mass. Specifically, this means that small amounts of additional mass must be added to the B-half of the wheel until the mass of both halves is equal. Alternately, one could choose to remove mass from the heavier A-half with a grinder, or by drilling holes to remove material. A balanced wheel will no longer favor a particular orientation, and that same wheel, spun at high speed, will run more smoothly and vibrate far less.

A more sophisticated way to measure the balance of a rotating machine is to do so while it is in motion. This is called dynamic balancing. The benefit of dynamic balancing is twofold. First, even tiny static imbalances become much more apparent, and second, the technique can reveal imbalances that are not detectable by static balancing at all.

If you've ever had the misfortune to have to visit a tire shop to have a tire fixed or replaced, you'll note that after a tire is mounted on a rim, it's almost always fitted with one or more metal weights before being reinstalled on your vehicle. These weights, which are attached to the wheel's rim with steel clips, are sized and positioned so as to cancel out vibration that arises from natural variation in wheel rims and in the tires themselves.

The balancing process starts with the temporary mounting of the wheel to a shaft protruding from the side of a balancing machine. An electric motor in the machine spins the wheel while sensors measure deflection in the shaft to which the wheel is attached. By measuring the amount of deflection, the machine can infer how much counterweight must be added to correct the imbalance. In order to determine where the counterweights should be attached, the machine records not only the magnitude of deflections in the wheel shaft, but when those deflections occurred. In other words, the machine is able to correspond deflections in the wheel shaft with the specific angular position of the wheel at a given instant.

It occurred to me that a tire balancing machine, scaled down to size, could be a very useful tool for detecting and correcting vibration in small, model machinery. I wondered if something simple could be constructed to give me this capability, and this train of thought inspired some experimentation. Figure 7-13 depicts a test fixture I knocked together in order to test out my ideas.

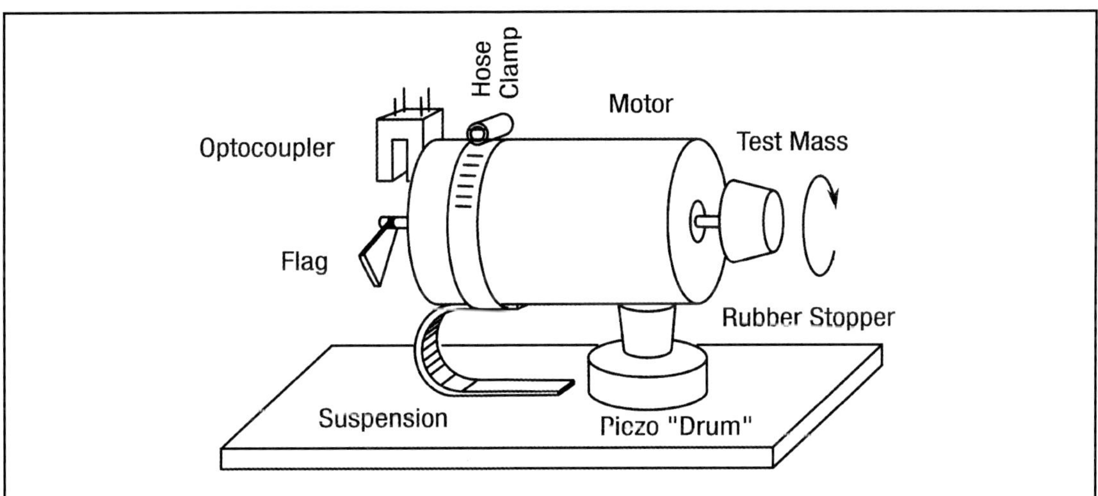

***Figure 7-13**: An experimental device to balance small rotating parts. A motor rotates a test mass. Vibration in the test object causes the motor to rock vertically, up and down. This motion is conveyed to a piezo "drum," which generates an electrical signal proportional to the displacement of the motor. An optocoupler and shaft flag provide timing pulses from which angle information can be derived.*

First, I found a small DC electric motor. There is nothing special about this motor, except that it has two shafts, one that protrudes from the front, and one that projects from the rear. On the rear shaft of the motor, I attached a plastic flag or vane which twirls about as the motor rotates. Epoxied to the back of the motor is an optocoupler, which in turn is connected to the optocoupler interface circuitry that I described earlier. Rotation of the motor shaft results in the generation of a series of pulses (review figure 7-9).

I used some strips of scrap brass stock to fashion a suspension/mount for the motor. The suspension is attached to the base of the instrument with screws and nuts, and to the rear-end of the motor with a hose clamp. The mount design is intended to allow the motor to flex upward and downward. At the same time, it is intended to discourage any side-to-side motion of the motor.

Toward the front of the motor, I built a deflection sensor which is comprised of a piezoelectric disk, a plastic ring, and a hard rubber stopper.

Piezo disks are often used as audible indicators ("speakers" if you will) in consumer electronics. When an electric current is applied to the disk, it deforms, causing it to emit audible noises. Depending upon the applied signal, these disks can be made to beep, click, chirp, or play low-fidelity music and speech. Piezo disks can be found in telephones, in electronic novelty greeting cards, in the backs of some electronic watches, electronic toys, and similar devices.

Now while it is true that an applied electric current can cause the disk to distort, the process is reversible. If the disk is mechanically distorted, an electrical signal is produced. I glued my piezo disk to a plastic support ring, which in turn, was glued to the base of my instrument beneath the front of the motor. In effect, the piezo disk is like the head of a drum. Distort the head and the piezo-electric effect will cause the generation of an electrical signal.

Between the piezo disk and the motor, I installed a rubber stopper. The stopper was glued to the center of the piezo disk with a drop of "super" glue, and attached to the motor with a plastic cable tie that passed through a hole in the stopper. The stopper conveys any up-and-down movement of the motor to the piezo disk.

Finally, to the front shaft of the motor, I attached a die-cast metal instrument knob. This component was completely arbitrary, and simply represents a test mass, a mechanical part to spin and to analyze. Not shown in figure 7-13 is a long bolt, which was threaded into the side of the knob, expressly for the purpose of rendering it asymmetrical and unbalanced.

In summary, this is what I expected the equipment just described to do:

- The motor, when powered up, would spin the test part, i.e., the die-cast metal knob.
- Since the knob is unbalanced, it would cause the motor to vibrate and shudder.
- The motor suspension mount would resist side-to-side motor movement, but easily flex in the vertical (up-down) axis.
- Vertical movement of the motor would be transferred to the piezo disk through the rubber stopper, causing it to generate an electrical signal.
- I expected the signal to be representative of the magnitude of vertical motor deflection due to the imbalance in the die-cast knob. The greater the vibration, the greater the signal produced by the piezo disk.
- The flag and optocoupler at the rear of the motor would generates a pulse, once per motor revolution.

The sound port on my laptop computer allows for stereo inputs. I fashioned a cable with a plug, wired the piezo disk to the right channel, and the optocoupler to the left. This allowed me to record both signals at the same time. Figure 7-14 is typical of the waveforms I recorded during my experiments. The top trace represents the signal coming from the optocoupler, the bottom trace comes from the piezo disk. Now let's consider, for a moment, what these squiggles might actually mean.

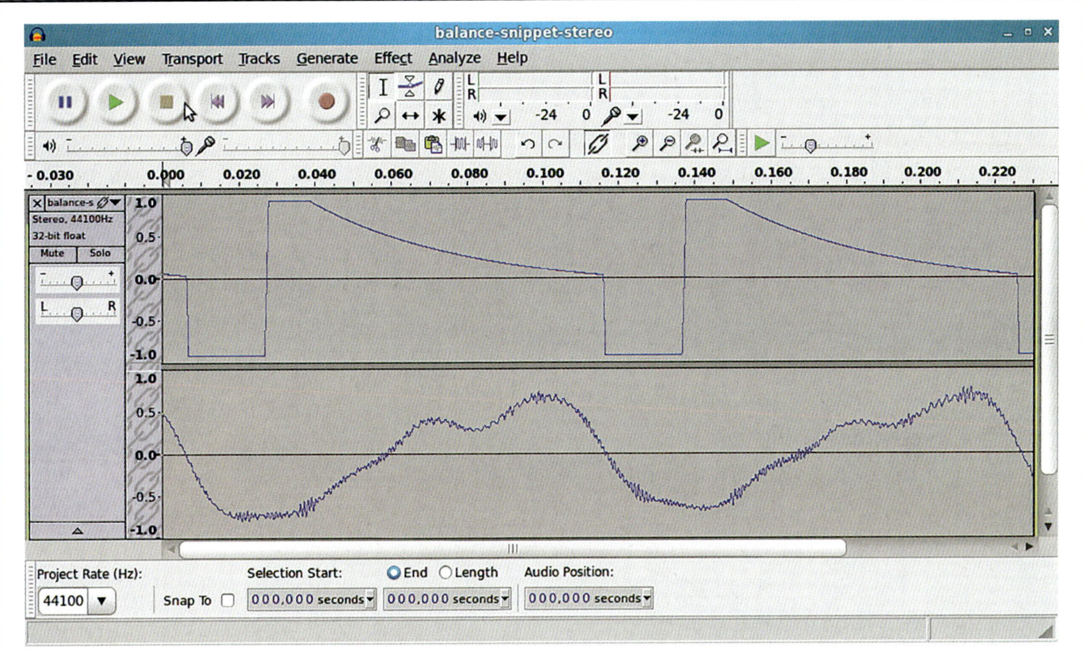

***Figure 7-14**: Data collected on an imbalanced test object attached to the machine depicted in figure 7-13. The top trace is generated by the optocoupler. The drop from zero to -1.0 signifies the start of a rotation. The bottom trace represents the magnitude of vibratory motion. The placement of artifacts in the bottom trace with respect to the start-of-rotation landmarks in the top trace provides information on angular displacement.*

The optocoupler trace (top) begins with a plunge from the zero level to a negative level, where it remains for a short interval. This transition represents the instant where the flag on the motor shaft has entered the space between the legs of the optocoupler.

The motor advances, and shortly thereafter, the flag moves out from between the optocoupler's legs. The signal shoots from negative to positive. The signal stays positive for a while and begins to droop toward zero, until the a full rotation of the motor's shaft has been completed. Then, the shaft flag enters optocoupler again, the trace goes negative, and the pattern is repeated (the drooping apparent in this signal is an artifact produced by the sound port's inability to handle frequency components below 20 Hz. For the purposes of the experiment, this behavior can be ignored).

The drop from zero to negative, that is to say, the point at which the vane just enters the optocoupler, is a useful reference point. I can arbitrarily define that as the start of every rotation. Since a full rotation spans 360 degrees, the span between successive pulses on the trace must represent 360 degrees.

It follows then that half the distance between successive pulses must represent 180 degrees. One-quarter the distance means 90 degrees, and so forth. Thus, the reference pulses created by the optocoupler and the time scale provided by the *Audacity* software serve to form a kind of protractor, with which shaft angle displacement can be deduced.

The signal produced by the piezo disk, depicted in the lower *Audacity* trace, shows evidence of repetition, which is precisely what we'd expect from an unbalanced load that is spinning. About 60 degrees after the start of the each rotation cycle, an imbalance in the test part causes the motor to jiggle and deform the piezo disk in the negative direction.

Logic dictates that an imbalance on the test part causing a "push" on the piezo disk at the 60 degree position, should then "pull" on the piezo disk when the test object has rotated a half turn. A half turn is 180 degrees. If we add 180 degrees to the initial 60, we would therefore expect to see a positive displacement in the graph somewhere near the 240-degree mark. If we study the data in figure 7-14, we see there is, in fact, a large positive excursion around 240 degrees. Granted, this excursion appears as a pair of "bumps" (an anomaly I'll comment on in just a second) but overall, the signals were consistent with what I would expect.

With additional experiments, I verified that the worse the imbalance of the rotating part, the greater the excursions recorded by the piezo disk. I also satisfied myself that, by studying the recorded signals, and using the optocoupler pulses as a tool to infer angular displacement, I was able to predict where counterweights should be placed in order to correct imbalances in the test part.

Even in its admittedly crude form, the device in figure 7-13 basically works, but there is obvious room for improvement.

If you've ever clamped a common ruler to the edge of desk with your palm and then plucked the free end, you've heard the "b-r-r-r-r" sound the ruler makes as it vibrates up and down. I think that something similar can happen with the springy motor mount in my instrument, and furthermore, I suspect that this accounts for the double "positive" bump in the piezo signal I recorded. The addition of damping might improve this, though I think it best to re-engineer to eliminate this type of suspension entirely.

In addition, because the test object is mounted directly to the motor shaft, there is no way to distinguish between vibration generated by the test object and vibration that might originate in the motor itself (the double bump could have come from the motor!). So, while we're eliminating the motor suspension, we should eliminate the motor, too.

Let us therefore re-imagine the balancing machine, starting with a simple steel shaft, perhaps 8-inches in length. At one end, we attach the test article, that is to say, the component we are interested in trying to balance.

At the other end of the shaft we affix a small pulley. The pulley is fitted with a rubber belt, which in turn is coupled to an electric motor. The motor provides the force to spin the test object.

The shaft is made to run on two pillow blocks with ball bearings. Instead of mounting the pillow blocks directly to the base of this new machine, they are instead mounted to the suspension depicted in figure 7-15. One such suspension assembly appears at each end of the shaft.

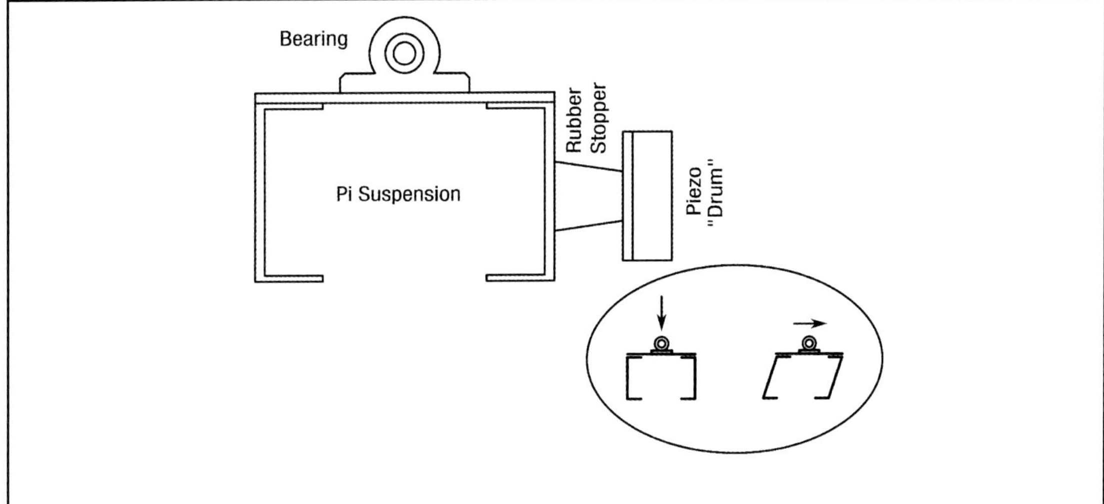

***Figure 7-15**: An alternative suspension for the balancing machine comprised of a pillow block and a pi-shaped support structure. Lateral vibration is communicated through a hard rubber stopper to the piezo drum sensor. The inset shows how this is possible: While the suspension does not yield to vertical force, lateral force can cause the suspension to shift or skew (the skew has been exaggerated here for clarity).*

The suspension is a pi-shaped structure with two legs, a top, and pillow block mounted to the top. Fabricated from aluminum strip, it is much more rigid than the prior design, and more-or-less unyielding in the up-and-down direction. However, its geometry allows for some lateral displacement, or skew, in response to unbalanced forces on the shaft. Adjacent the suspension, off to one side, we mount our piezo sensor, which is coupled to one of the suspension's legs.

As before, shaft position can be tracked with an optocoupler and flag attached to the shaft itself. Alternately, a piece of reflective tape could be affixed to the pulley, and a reflective optical sensor used in lieu of the optocoupler.

An observant reader might note that whereas the first design detected vibration on the vertical axis, the pi-shaped suspension and sensor measures lateral vibration. This difference is of no real consequence, so long as we can determine associate shaft angle with specific vibrational events. The optocoupler provides that information.

I would expect that the comparative rigidity of the pi-suspension would eliminate the bounce observed in the looser, more springy design. The electric motor, driving the machinery through a belt, is effectively isolated from the shaft carrying the test object. Any imbalance present in the motor itself should therefore have little or no effect on measurements made of the test object.

A final comment; the piezo disk's sensitivity is not limited to large low-frequency excursions, caused by deflection of a suspension.

Piezo disks are quite effective as microphones, which means that a signal recorded from one can be cluttered with all sorts of irrelevant noise, including motor whine, ambient noises, and even speech. Much of this racket can be filtered away with a low-pass audio filter. A circuit of this type is easy enough to build, though I avoided having fabricated this circuitry by using the software-based frequency equalizers that are built into *Audacity*.

Conclusion

While I hope you've found the preceding pages interesting reading, they will have achieved their ultimate purpose only if you take the next steps on your own. Reading and doing are two very different things. There is no substitution for working with your own hands, and no greater reward than the pleasure that comes from the transformation of an idea into something physical and tangible.

That's why, in closing, I invite you to visit a church sale, second-hand store, or flea market — and don't be afraid to rummage through the trash. Collect that hank of wire, the odd bearing, or that piece of scrap metal. Contemplate the treasures you find, daydream, and doodle. Find value in what others cannot, and be empowered by the tools that you have, not limited by the ones you don't.

What will the next "Marvelous Magnetic Machine" look like? That question is yours to answer.

Appendix: Open Source Software

It is safe to say that computers have invaded the practice of every discipline or field of interest you can imagine. Powerful software can be purchased to accomplish a breathtaking variety of tasks. The problem with commercial software is that, often times, the price tag and licensing conditions are equally breathtaking, so much so as to discourage that software's use.

A comparative handful of commercial players dominate the market, and I suspect that this domination has lead many end-users to think that no other options exist. Not true!

Over the last 20 years, the so-called "open source" software movement has revolutionized the computer world. The open-source idea has inspired thousands of individuals from around the world to volunteer their time and talents in the creation of absolutely fabulous pieces of computer software. Despite its grass-roots origin, much of this software meets (and sometimes exceeds) the quality and features found in corresponding commercial products.

Open source software is "free" in the sense that the developers don't demand payment as a condition for downloading and using it. Perhaps more important, an open source license grants the end-user literal access to source code — the feedstock from which application programs are compiled and built. That means if you are so inclined, you can enhance, modify, or customize applications as you see fit.

I made extensive use of open-source software in the production of this book's content, manuscript, and illustrations. My text was prepared with *LibreOffice*, a full, open-source, Microsoft-compliant office suit. Simple line illustrations were prepared with *Inkscape*, a powerful vector graphics drawing program. My photographs were cropped and processed with *GIMP*, the Gnu Image Manipulation Program. The number of minor tools and utilities that I used are too numerous to list.

Worthy of special note is *FreeCAD*, the design package I used to generate the three-dimensional mechanical models that appear throughout this book. Using *FreeCAD*, you can design mechanical parts, edit them, manipulate them, and assemble them to test for fit. As evidenced by the illustrations in this book, assemblies can be quite complex. It's a great tool for exploring ideas without having to consume physical materials or do physical machining work.

By the way… if you own a 3-D printer, you're basically just a few mouse clicks away from exporting *FreeCAD* objects to physical reality.

Another interesting software package is called *Audacity*. It's the audio recording/editing package that was used to capture, process, and plot motor signals, as described in Chapter 7. Though intended for music and speech, it contains a number of powerful built-in functions that allow filtering and even spectral analysis of recorded signals.

The software packages I mentioned are supported by vibrant communities of users and enthusiasts who frequent various online forums. I've never encountered a software problem and then failed to find someone on the Internet who could answer a question, offer advice, or propose a work-around. Other users have produced endless numbers of software tutorials on YouTube. Video tutorials can be a great way to become proficient with unfamiliar packages very quickly.

In my case, all of the above software was run on laptop computers on which my favorite flavor of *Linux* operating system was installed. *Linux Mint*, another gem in the crown of open-source, is a relatively lean operating system, stable, fast, and installs with a software package manager that gives you single-click access to tens of thousands of additional free and open-source software packages.

If, however, Microsoft Windows© is your cup of tea, fear not. Virtually all of the open-source applications I've mentioned above offer versions that will run natively on Windows.

I applaud the work of the open-source software community, and I encourage you to explore what this powerful alternative has to offer.

Some Software Packages Used in the Preparation of this Book

Software	**Application**	**URL**
Linux Mint	Operating system (alternative to Microsoft Windows)	linuxmint.com
Inkscape	Vector graphics creator/editor	inkscape.org
LibreOffice	Office suite including word a processor, spreadsheet, etc (Alternative to Microsoft Office)	www.libreoffice.org
Audacity	Audio recording, editing/processing, and playback	audacityteam.org
GIMP	Bitmap image editing/processing (Alternative to Adobe Photoshop)	www.gimp.org
FreeCAD	3-D modelling software	freecadweb.org

Appendix: Source Material and Interesting References

ARRL, *The ARRL Handbook for Radio Communications* 6-Volume Set. Newington, CT:ARRL, 2019

Brown, William Fuller, "Chapter 8 Magnetic Materials." *Handbook of Physics*. New York: McGraw-Hill Book Company, 1967: pages 4-dash-142 to 4-dash-143 (relative μ of various materials)

Graham, Frank D., Vol. 1 of *Audels New Electric Library*. New York: Theo. Audel & Co., Publishers, 1948: page 175 (Oersted's discovery), page 177 (right-hand rule), page 185 (solenoid right-hand rule)

Gregory, Frederick, *Episodes in Romantic Science*, 1998 (accounts of Hans Christian Oersted's work)
http://www.clas.ufl.edu/users/fgregory/oersted.htm

Graham, Kennard C., *Fundamentals of Electricity*. Chicago: American Technical Society, 1965

Herndon, J.M., "*Nuclear Georeactor Generation of the Earth's Geomagnetic Field.*" Current Science. Vol. 93, No. 11. 10 December 2007: pages 1485-1487 (a newer theory of the origin of the earth's magnetic field)

Horowitz, Paul, and Hill, Winfield., *The Art of Electronics*. Cambridge: Cambridge University Press, 1989

Morgan, Alfred, *The Boy's Book of Engines, Motors and Turbines*, New York: Charles Scribner's Sons, 1946. pages 221-233 (plan for simple homemade electric motor)

Phillips, Tony, *Earth's Inconstant Magnetic Field*, Dec 2003 (origin of the earth's magnetic field)
<http://www.nasa.gov/vision/earth/lookingatearth/29dec_magneticfield.html>

Tektronix. *Primer XYZs of Oscilloscopes*. Beaverton, OR: Tektronix, 2000

Underhill, Charles N., *Solenoids Electromagnets and Electromagnetic Windings*, D. Van Nostrand Company, Inc., 1921. page 124 (pull force vs air gap for coned plungers with different head angles)

Underhill, Charles N., *Solenoids Electromagnets and Electromagnetic Windings*, D. Van Nostrand Company, Inc., 1921. page 135 (pull force vs air gap for coils with external armatures)

Unesco, *700 Science Experiments for Everyone*, Doubleday & Company, Inc., 1962. pages 182-183 (three plans for primitive homemade electric motors)

U.S. Navy, *Navy Electricity and Electronics Training Series – Module 1 – Introduction to Matter, Energy, and Direct Current* – NAVEDTRA 14173. United States of America: U.S. Navy, 1998

U.S. Navy, *Navy Electricity and Electronics Training Series – Module 2 – Introduction to Alternating Current and Transformers* – NAVEDTRA 14174. United States of America: U.S. Navy, 1998

U.S. Navy, *Navy Electricity and Electronics Training Series – Module 3 – Introduction to Circuit Protection, Control, and Measurement* – NAVEDTRA 14175. United States of America: U.S. Navy, 1998

U.S. Navy, *Navy Electricity and Electronics Training Series – Module 4 – Introduction to Electrical Conductors, Wiring Techniques, and Schematic Reading* – NAVEDTRA 14176. United States of America: U.S. Navy, 1998

White, Harvey E., *Modern College Physics*, D. Van Nostrand Company, Inc., 1956. page 57 (flux density inside of a solenoid)

White, Harvey E., *Modern College Physics*, D. Van Nostrand Company, Inc., 1956. page 521 (formula to compute force exerted between two magnetic poles)

Windsor, H. H., *The Boy Mechanic Volume I*. Chicago: Popular Mechanics Co. Publishers, 1913

Author's Biography

H.P. Friedrichs is a degreed electrical engineer (BSEE), inventor, and author with more than three decades of experience working in domains ranging from audio, medical, and radio, to software, automotive, and aerospace. At present, he is a Principal Engineer with Honeywell, involved in the design and support of specialized equipment used for testing and validating aircraft power generation products.

He has five U.S. patents to his credit and holds three radio licenses including Extra-Class Amateur (AC7ZL), Commercial Radio Operator with Radar Endorsement and GMDSS Operator/Maintainer with Radar Endorsement. He is also a certified VE.

Friedrichs is the author of numerous technical articles appearing in a variety of magazines, newsletters, and web sites but is best known for his books *The Voice of the Crystal* and *Instruments of Amplification*. Now cult classics among "from-scratch" electronics experimenters, these books have enjoyed favorable reviews from the editors of such prestigious periodicals as *QST*, *CQ Magazine*, *Practical Wireless*, and *Make Magazine*.

H.P. Friedrichs lives in Tucson, Arizona, with his wife and his German Shepherd/ laboratory assistant — who is prone to "borrow" books and tools but not return them.

The author's website is: **www.hpfriedrichs.com**.

Also Available from ArtisanIdeas.com:

The Voice of the Crystal: How to Build Working Radio Receiver Components Entirely from Scratch

By H.P. Friedrichs

"*The Voice of the Crystal has a stellar reputation amongst radio geeks, and for good reason. This guide to building radios is filled with great ideas and utterly awesome projects.*" - **Gareth Branwyn**, *Make Magazine.*

To many, crystal radio construction means purchasing electronic components. However, the very best tinkerers can build their own! In *The Voice Of The Crystal*, author H. P. Friedrichs reveals how to construct all of the necessary components "from scratch".

Inside this volume you will find 185 pages of practical information on the fabrication of electronic components suitable for use in building crystal radio sets. Basic theory and simple analysis is combined with dozens of examples of historical practice, work by contemporary experimenters, and construction details for many instruments fabricated by the author himself.

Inside are discussions and plans for three different homemade headphones (including one fabricated from cigarette lighter parts), detectors, fixed capacitors, a rotary variable condenser, single layer, spider web and basket coils, and a precision double-slider coil. Throughout, the author shares his thinking and practical experience in the construction of these devices. *The Voice Of The Crystal* is profusely illustrated, containing well over 120 photos and hand-prepared illustrations.

Also Available from ArtisanIdeas.com:

Instruments of Amplification:

Fun with Homemade Tubes, Transistors, and More

By H.P. Friedrichs

Instruments of Amplification, written and illustrated by H. P. Friedrichs, is jam-packed with nearly 300 pages of history, science background, basic theory, and hard-to-find hands-on details pertaining to the construction of an amazing array of homebrew amplifying devices.

Rooted in the same "build-it-from-scratch" philosophy that made his first book, *The Voice of the Crystal*, a success, *Instruments of Amplification* reduces complex devices to their essential elements and then shows how they can be constructed from commonly available materials.

In the process of building, you'll also learn secrets that will find application to other projects. Learn to drill a hole in glass, generate high voltages, or create and measure a high vacuum. Learn how to dismantle a lightbulb, harvest carbon from old batteries, or deposit a layer of metal onto glass so thin that it is transparent! How about creating your own primitive semiconductor materials from garden-shed chemicals? The list goes on and on!

The wealth of information contained in *Instruments of Amplification* is augmented by 150 photos, illustrations, and engravings, in addition to numerous charts, tables, and formulas. Readers interested in further exploration will appreciate the 120+ references to period books, magazines, CD-ROMs and websites.